Trout Farming Handbook

SIXTH EDITION

Stephen Drummond Sedgwick

FISHING NEWS BOOKS

Fishing News Books
A division of Blackwell Science Ltd
Editorial Offices:
Osney Mead, Oxford OX2 0EL
25 John Street, London WC1N 2BL
23 Ainslie Place, Edinburgh EH3 6AJ
238 Main Street, Cambridge
  MA 02142, USA
54 University Street, Carlton,
  Victoria 3053, Australia

Other Editorial Offices:
Arnette Blackwell SA
  1, rue de Lille, 75007 Paris
  France

Blackwell Wissenschafts-Verlag
  GmbH
  Kurfürstendamm 57
  10707 Berlin, Germany

Feldgasse 13, A-1238 Wien
Austria

First edition published by Fishing
  News Books 1973
Second edition published 1976
Third edition published 1978
Fourth edition published 1985
Fifth edition published by Fishing
  News Books, a division of
  Blackwell Science, 1990
Sixth edition published 1995

Printed and bound in Great Britain
  at the Alden Press Limited,
  Oxford and Northampton

DISTRIBUTORS

  Marston Book Services Ltd
  PO Box 87
  Oxford OX2 0DT
  (*Orders:* Tel: 01865 206206
            Fax: 01865 721205
            Telex: 83355 MEDBOK G)

USA
  Blackwell Science, Inc
  238 Main Street
  Cambridge, MA 02142
  (*Orders:* Tel: 800 215-1000
            617 876-7000
            Fax: 617 492-5263)

Canada
  Oxford University Press
  70 Wynford Drive
  Don Mills
  Ontario M3C 1J9
  (*Orders:* Tel: 416 441-2941)

Australia
  Blackwell Science Pty Ltd
  54 University Street
  Carlton, Victoria 3053
  (*Orders:* Tel: 03 347-0300
            Fax: 03 349-3016)

A catalogue record for this book is
available from the British Library

ISBN 0–85238–232–4

Contents

List of illustrations and tables   vii

Introduction   xi

1 The farmer's fish   1
Rainbow trout. Brown or German trout. Arctic char

2 An outline of farming methods   9
Water supply. Pond culture – the Danish system. Raceways.
Fish tanks. Fish cages. Shore enclosures

3 Brood stock and egg production   19
Selection for quick growth. Selection for egg size. Mixing brood
stock. Cryogenic storage. Sex reversal and induced sterility.
Buying eggs. Cross breeding. Working with brood stock.
Hatcheries. Incubation. Hatchery design

4 Fingerling production   37
Fry rearing. Production methods. Water supply. Flows and fish
density. Tank covers. Indoor tanks. Sale and delivery

5 Trout culture in earth ponds   45
Pond construction. Screens. Feeding in earth ponds. Grading.
Stocking and flows. Aeration. Cleaning ponds. Effluent from
earth ponds. Market production

6 Raceways   59
Design and construction. Flow and fish density. Feeding and
grading. Disease risk. Investment

7 Tank farming   63
Water supply. Fish density and flow. Tank farm construction.
Low-head recycling

8 Cages in freshwater   71
Sites. Anchorages. Collars and cages. Servicing. Shore bases.
Fish feeding. Grading and killing. Freshwater cage farm
production

9 Rainbow trout farming in salt water   77
Saltwater tolerance. Acclimatization to sea water. Shore farms.

Tidal enclosures. Sea cages. Anchorages and shore bases. Cages
on a shore walkway. Floating 'islands'. Fish husbandry.
Maintenance. Sea fattening. Osmotic stress. Steelhead or sea-
going rainbow trout. Steelhead culture. Recirculated water

10   Trout farm design, equipment and operation   91
Sites. Operational equipment. Hatchery equipment. Heating
water. Fish handling. Feeding. Security

11   Trout food and feeding   101
Basic diet. Calorie requirements of trout. 'Dry' feeds for
rainbow trout. Quantities of food and conversion rates. Moist
pellets. Hygroscopic pellets. Fish silage. Times of feeding.
Growth in fresh water. Sexual maturity. Flesh colour of rainbow
trout. Dietary deficiencies. Controlling growth. Feed cost

**12   Common diseases – recognition, treatment and
prevention**   113
Bacterial diseases. Protozoan diseases. Diseases caused by
flukes. Virus diseases. Diagnosis. Other afflictions. Disinfection.
Immunization. Some useful chemicals. Hygiene

13   Pollution from fish farms   135
Causes of pollution. Pollution load. Danger to the environment.
Danger to other fish farmers. Earth ponds. Tank farms and
raceways. Cage farms. Treatment

14   Processing trout   143
Handling. Slaughter. Fish packing in ice. Freezing and frozen
cold storage. Types of freezer. Packaging frozen fish. Cold
storage. Gutting. Filleting. Smoking. Hot smoking. Cold
smoking. Dangers in cold-smoked trout. Other methods. Design
and operation of processing plant

15   Markets, profits and losses   155
Profit margin. Running costs of trout farms. Growth rate. Trout
for the market. Short-falls in production. Fish losses. Insurance.
Specialist production. Co-operatives. Is it worthwhile?

Index   163

Illustrations and tables

Illustrations

1.1	The farmer's fish (rainbow trout, Arctic char, brown trout, sea trout)	3
1.2	Brown trout	4
1.3	Trout − external anatomy	4
1.4	Trout viscera	5
2.1	Danish rainbow trout farm	12
2.2	Raceways in S.W. France	13
2.3	Concrete tanks in Sweden	15
2.4	The Kames floating fish cage	16
2.5	Shore enclosure	17
3.1	Selected brood stock	20
3.2	Rainbow trout × Arctic char	25
3.3	Development − egg to fry	26
3.4	Hatching troughs	29
3.5	Trout eggs	31
3.6	Small hatchery plan	33
3.7	Hatchery constructional design	34
4.1	Fry farm	39
4.2	Concrete fry tanks	40
4.3	Site plan of fry rearing unit	41
4.4	Fry tanks	42
4.5	Water supply and drainage to fry tanks	43
5.1	Layout plan for earth ponds	46
5.2	Earth ponds	47
5.3	Inlet channel and back channel	48
5.4	'Monk' pond outlet	49
5.5	Hand feeding	50
5.6	Demand feeder	51
5.7	Gun feeder	52
5.8	Sorting pontoon	53
5.9	Grading tanks	54
5.10	Pond liner	56

5.11	'Portion' size rainbow trout	57
6.1	Aerators and mobile fish feeder	60
6.2	Fish pump and grader	61
6.3	Large raceway farm	62
7.1	Grp tank − wall section	63
7.2	Grp tank sunk on site	64
7.3	Tank base	64
7.4	Galvanized-iron tanks	65
7.5	Low-cost intake valve	66
7.6	Intakes to grp tanks	66
7.7	Tank outlets	67
7.8	Layout for low-head recycled water supply	69
7.9	Danish tank farm with low-head recycling	69
8.1	Freshwater cage site	72
8.2	Landing stage at cage site	73
8.3	Fish feeders on fish cages	74
9.1	Adaptation to seawater	79
9.2	Saltwater shore farm	80
9.3	Sea cages	82
9.4	Sea cages	83
9.5	Floating 'island'	84
9.6	Steelhead trout	87
9.7	Recirculation system	89
10.1	Intake screen	93
10.2	Pond outlet screen	94
10.3	Tank outlet screen	95
10.4	Warming hatchery water	96
10.5	Heat pump for warming water	97
10.6	Fish-pump and grader	98
10.7	Fry-feeders	99
12.1	*Costia*	118
12.2	*Octomitus*	120
12.3	*Gyrodactylus*	121
12.4	*Diplostomum* life cycle	123
12.5	VHS	126
13.1	Sludge separator	139
13.2	Oxygen injector	140
14.1	Quick-freezing	145

14.2 Packaging trout 147
14.3 Hot-smoking kiln 150
14.4 Trout steaks 152

Tables
2.1 Dissolved oxygen – parts per 100 000 10
3.1 Fish anaesthetics 28
9.1 Average growth rate 88
11.1 Minimum amino-acid requirements for salmonids 102
11.2 Daily vitamin requirements 103
11.3 Additional vitamin requirements 103
12.1 Some useful chemicals for treating fish 132
13.1 Pollution load per tonne of fish production 136
13.2 Ammonia contents 137
13.3 Percentage of un-ionized ammonia in seawater at different temperatures, salinities and pHs 142
14.1 Brining time for hot smoking 149
15.1 Percentages of running costs 157

Introduction

A good many people who were initiated into trout farming through earlier editions of this book may continue to find their original copies useful as a day-to-day guide. It can still be a good place to look up and be reminded of essentials. Most of that earlier information is still relevant but some useful fresh ideas have come along in the meantime and have been included in this sixth edition of the *Trout Farming Handbook*. Even established trout farmers who now grow tonnes of fish, as well as newcomers to the industry, need to be aware of these more recent developments.

More trout than ever is now on sale fresh, frozen and processed in all shapes and sizes. Customers have acquired a more selective taste that can influence their choice, but trout farmers must still get their fish to market at the right time and at the right price. These remain the simple essentials for success and still have to be achieved when faced by increasing opposition intended to protect the environment. An English trout farmer attracted adverse attention when large rainbow trout, which had escaped from his farm, were supposed to have spoiled the fishing for brown trout in a local river.

Rainbow trout were the first member of the salmon family to be farmed for food, but hatching and growing brown trout for restocking angling waters had been going on in the British Isles for close on a hundred years before rainbows came on the scene. Now some pond farmers, coping with a lack of water and other restrictions that prevent them from producing enough fish for the table market, have gone over or returned to growing fish for angling. They are stocking small angling ponds they have excavated for themselves, some as small as a quarter of an acre. This has proved a profitable enterprise provided the trout farm and the fishing ponds are not too far from an urban area. The secret is that the anglers who pay to fish must then buy the trout they catch. The trout are weighed and costed at only slightly less than the retail price in the local fishmongers.

1 The farmer's fish

Rainbow trout
(Salmo
gairdneri)

The home range of this species extends from the Kuskokwim River in Alaska south through British Columbia to Baja in California. It is primarily a native of the coastal rivers of western North America but also occurs on the eastern side of the Great Divide in the headwaters of the Peace River in British Columbia and the Athabasca in Alberta. There is also a native population in the Rio Casa Grandes in the Mexican province of Chihuahua. The migratory race of sea-going steelhead grow more quickly and are generally bigger than the non-migratory race which lives out its life cycle in rivers but the largest rainbow trout are found in freshwater lakes. A near relative is the Far Eastern trout (*Salmo mykiss*) which occurs in the Kamchatka rivers in Asia.

There are two main varieties of rainbow trout, a sea-going form known as the steelhead, which occurs in most of the rivers draining into the Pacific Ocean, and a form permanently resident in fresh water. The freshwater form is subdivided into a number of racial types. The steelhead has a particularly rapid growth rate in salt water, equal in fact to that of the fastest growing Pacific and Atlantic salmon, and it can reach a weight of 7−10 kg (15−20 lb) after three years feeding in the sea. The freshwater form is slower growing but can attain a weight of 4.5 kg (10 lb) or more in favourable circumstances. When rainbow trout were introduced into high lakes in the South American Andes, which contain rich supplies of fish-food animals, they grew to weights in excess of 9 kg (20 lb). The sea-going and freshwater forms are quite distinct in external shape and are easily recognizable even at an early stage of life. The sea-going form is longer and slimmer than the freshwater variety.

The spawning season of the different races can be any time from September to April in the northern hemisphere, extending through the autumn and winter months on into the spring.

This applies to both sea-going and freshwater types. Rainbow trout transplanted to the southern hemisphere retain the same seasons for spawning and this provides an opportunity to obtain eggs in practically any month of the year.

Rainbow trout (Fig. 1.1) are the fish farmer's fish and have been domesticated and cultured for the table market since the late nineteenth century. They now form the basis of an industry which has developed and continues to grow in importance in practically every country which can provide a suitable fresh or saltwater environment. European imports have consisted almost entirely of eggs from parent fish of the freshwater type, taken from brood stock which has been bred in captivity. European producers of rainbow trout eggs now rely almost entirely on domestic stocks. Most of the eggs generally obtainable from breeders in Europe or North America come from brood fish which are descended from a mixture of spring and autumn spawning fish. The spawning times of particular brood stocks have been stabilized and commercial producers can provide eggs from early, middle or late-spawning parent fish.

Brown or German trout (*Salmo trutta*) The most widely distributed of the world's trout, this species occurs naturally in many different, racially distinct forms throughout Europe, parts of North Africa and the Middle East and the western side of Asia. The most familiar form is perhaps the typical river trout of western Europe (Fig. 1.2– 1.4). They can remain dwarfed in size, never reaching a length of more than 20–30 cm and mature early when they are only two or three years old. In rich chalk streams with alkaline water and a plentiful food supply river trout can grow to a weight of 2–5 kg becoming sexually mature after three to four years and surviving for 10–14 years. Fish in small lakes will ascend any tributary stream to spawn and the young fish feed and grow to sexual maturity in the lakes. Rich, eutrophic lakes will usually support a population of fast-growing fish. Oligotrophic lakes producing little in the way of food fauna but also having little available spawning ground may still have a small population of fairly large, slow-growing trout. On the other hand, lakes of this type with

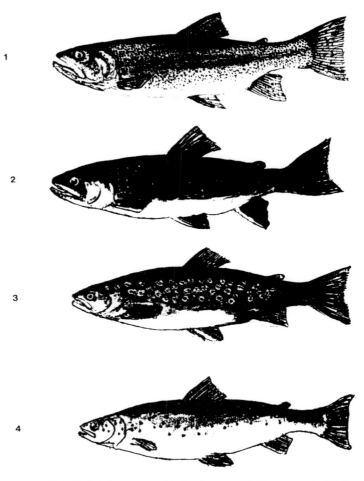

Fig. 1.1 (1) Rainbow trout; (2) Arctic char; (3) Brown trout; (4) Sea trout

good spawning ground in tributary streams will have a large population of small fish.

Salmo trutta in the largest lakes which are virtually inland seas may adopt a completely different behaviour pattern compared to other members of this species. The huge lakes

Fig. 1.2 Brown trout (*Salmo trutta*)

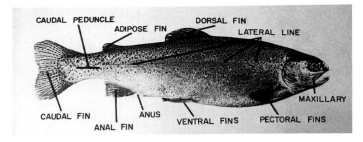

Fig. 1.3 Trout — external anatomy

where this racially distinct type of trout is to be found are usually on big rivers. The trout behave very much like Atlantic salmon and eventually grow to comparable size although they take longer to do so and spawn a number of times. The environmental factor which appears to induce this lifestyle is the presence in the lake of a large population of some food fish, usually a corregonid or whitefish. They are fully migratory and usually spawn in the main river flowing into the lake. Although they may not be ripe to spawn until late September and October, the adult fish may enter the river as early as June or July and run upstream for 50 miles or more before reaching their spawning grounds. The young fish spend one to

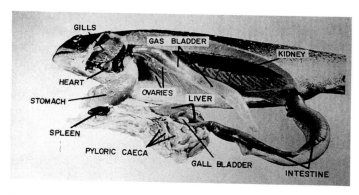

Fig. 1.4 Trout viscera

three years of parr life in the river before migrating downstream to the lake. The growth rate is quite comparable to that of the anadromous race and the fish can reach a weight of more than 15 kg.

The anadromous, sea-going form of *Salmo trutta* occurs in European rivers from the White Sea to the north coast of Spain. Related forms migrate to and from the rivers running into the Black and Caspian Seas. The seaward migration of young fish usually takes place in the spring or early summer when they have reached a length of 15−25 cm and an age of one to five years, taking longer to reach migratory size in the more northern rivers. They do not range far to sea usually remaining well inside the continental shelf during their marine life.

Some immature or precocious fish return to fresh water after having spent only a few months at sea. Most fish return after having spent one to three years at sea and having reached a weight of 1−2 kg, but some spend up to five years marine feeding and grow to a weight of 7−8 kg before returning on first spawning migration.

Sea trout can survive to spawn many times in either successive or alternate years, returning to sea to feed in the interim.

Spawning takes place in October/November and in northern rivers may take place after Atlantic salmon have finished spawning.

The largest anadromous form of *Salmo trutta* is the Caspian race. Fish of over 45 kg are occasionally taken while on spawning migration into rivers in the USSR and a monster of 50 kg was reported from the Kura River in Azerbaijan.

The Arctic char (*Salvelinus alpinus*) This is the most widely distributed species of char. It is present with racial variations in most of the rivers and lakes in the more northern land areas round the Pole. It is primarily an Arctic species and the anadromous races occur only in the far north of Europe, Asia and North America, including rivers in Greenland. Fishing for the sea-going races was of basic, economic importance in the lives of the native Eskimaux and Yakut peoples. The completely freshwater form exists mainly at the southerly end of the range of the species and there are numerous racially distinct relict varieties in lakes throughout the sub-Arctic and temperate zones in the northern hemisphere. The individual characteristics of the separate stocks have developed through isolation since the last Ice Age. Both the non-migratory and migratory forms can occur in the northern part of the range of the species.

The life cycle of the sea-going race is fairly constant throughout the range of the species. There are two forms, one of which returns to spawn in the shallow waters of lakes while the other type remains in running water and spawns there. The spawning season is September to October. The female fish produces 3000–4000 eggs per kilogram of body weight which measure 3–4 mm in diameter. The lake-spawning variety does not make a redd or nest but spawns over rough, stony ground where the eggs lie between the stones until they hatch. The kelts or spawned-out fish return to sea on the spring floods after the ice melts in late May or June of the following year.

The newly hatched fry measure 15 mm in length and have a large yolk sac which is absorbed in about 30 days. The young fish migrate to sea after a variable period of parr life, some of which may be spent in brackish water.

The marine food consists mainly of small fish and the growth rate varies considerably between different stocks. The fish may take three to six years to become sexually mature reaching a weight of 1—3 kg.

The static population in lakes belong to one of several different basic types which can be either autumn- or spring-spawning and vary greatly in growth rate and eventual adult size. More than one of these basic types can exist in the same lake. The commonest type is relatively slow growing and feeds on invertebrates, either planktonic crustacea, or small bottom fauna in comparatively shallow water. A less common race lives in deep water on a pisciverous diet, growing rapidly to reach a weight of 1—2 kg at sexual maturity and continuing to grow-on so that the largest individuals eventually attain weights of 8—10 kg.

The quick-growing, pisciverous, non-migratory race and the sea-going race are of interest to fish farmers. They are hardy, easy to domesticate and grow comparatively quickly in captivity. The pisciverous race has an ability to retain carotin and quickly becomes red-fleshed. The species, either on its own or hybridized with another salmonid, seems to have considerable potential as a farmed fish.

2 An outline of farming methods

The most difficult obstacle in the way of the aspiring rainbow trout farmer is finding and acquiring a satisfactory site for a fish farm. The two basic essentials for this are that there must be a sufficient supply of clean water, of the right chemical quality, not only to provide for initial development but to allow for expansion; and also that the site must be reasonably accessible. A farm that is a long way from markets and from sources of food for the fish may too easily become uneconomic to operate because of the rising cost of transport, labour and other overheads.

Water supply The quantity of live fish held on a fish farm is normally calculated in tonnes. A freshwater rainbow trout farm, using surface water subject to the fluctuations in temperature typical in the temperate zone, must have an available water supply of 5 l/s per tonne of fish. The maximum water supply will be required when the greatest weight of fish is likely to be on the unit. This is during the summer months, before the main slaughter period, when water temperatures can also be ex-pected to be at a maximum. Progressively less throughput of water will be needed at lower temperatures. For example, a unit with water that could be guaranteed never to rise above 15°C could operate on a water supply of 3 l/s. These quantities are based on the assumption that the oxygen dissolved in the water is at saturation level.

Water chemistry It is essential that the water supply for a fish farm is free from pollution. The oxygen content should be 100% of saturation (Table 2.1). A neutral or mildly alkaline water is to be preferred with a pH of 7.0–7.5. A pH of less than 6.0 should be avoided, and it is particularly important to make sure that the pH does not fall below this level following periods of rainfall, when a surface water supply is used. The best water supply is

9

Table 2.1 Dissolved oxygen – parts per 100 000

Temp.	Percentage of saturation					
(°C)	100%	90%	80%	70%	60%	50%
1	1.43	1.29	1.14	1.00	0.85	0.71
2	1.39	1.26	1.12	0.98	0.83	0.70
3	1.36	1.22	1.09	0.95	0.81	0.68
4	1.32	1.19	1.06	0.93	0.79	0.66
5	1.29	1.16	1.03	0.90	0.77	0.64
6	1.26	1.13	1.01	0.88	0.75	0.63
7	1.23	1.10	0.98	0.86	0.74	0.61
8	1.20	1.08	0.96	0.84	0.72	0.60
9	1.17	1.05	0.94	0.82	0.70	0.58
10	1.14	1.03	0.92	0.80	0.68	0.57
11	1.12	1.01	0.90	0.78	0.67	0.56
12	1.10	0.98	0.88	0.77	0.66	0.55
13	1.07	0.96	0.86	0.75	0.64	0.53
14	1.05	0.94	0.84	0.74	0.63	0.52
15	1.03	0.93	0.82	0.72	0.62	0.51
16	1.01	0.91	0.81	0.71	0.61	0.50
17	0.99	0.89	0.79	0.69	0.59	0.495
18	0.98	0.88	0.78	0.68	0.58	0.49
19	0.96	0.86	0.77	0.67	0.57	0.48
20	0.94	0.85	0.75	0.66	0.56	0.47
21	0.92	0.83	0.74	0.65	0.55	0.46
22	0.91	0.82	0.72	0.63	0.54	0.45
23	0.89	0.80	0.71	0.62	0.53	0.445
24	0.87	0.79	0.70	0.61	0.52	0.44
25	0.86	0.78	0.69	0.60	0.515	0.43
26	0.85	0.76	0.68	0.59	0.51	0.42
27	0.83	0.76	0.67	0.58	0.50	0.415

one which has a stable pH, buffered by the presence of chalk in the water.

A proper chemical analysis must be made of the proposed water supply to any fish farm. This is the most vital factor toward success and neglect of it will inevitably lead to failure. Samples must be taken under all weather conditions and at all seasons. It is not enough to take them only on one day and it cannot be overemphasized how necessary it is to make quite certain that the chemistry of the water remains suitable for fish production at all times.

Silt in a surface water supply, at times of high flow following rain, can be a nuisance, but is seldom sufficient to kill the larger fish. It need not be regarded as a complete obstacle to development, but normally silt-free water is considered essential for a successful commercial fish hatchery. If spring water is used on a fish farm (or water from a source which does not have a natural fish population), it should be carefully tested for the presence of metal salts which can be toxic to fish, even at very great dilution. Spring water should also be tested to make sure that it is not super-saturated with air (nitrogen), as this can give rise to a condition known as 'gas-bubble disease', which is caused by the formation of bubbles of nitrogen in the tissues and to which young fish are particularly susceptible.

Water temperature The ideal water temperature for rainbow trout production is one that does not rise too high in summer nor fall too low in winter. A temperature of 18°C is regarded experimentally as being the optimum for rainbow trout metabolism. This means that it is the temperature at which rainbow trout make the best use of food from the point of view of the trout farmer, and it is the temperature at which the trout make the greatest conversion of food to tissue, in terms of both time and weight gain. So far it has not proved practicable to maintain a constant water temperature in the rearing and fish-fattening ponds on a fish farm, although this might theoretically be possible using cooling water from some source such as a power station.

It must be borne in mind that the higher the temperature of the water, the less will be its oxygen-carrying capacity. This means that proportionally fewer fish can be maintained in a given throughput of water, or alternatively more water must be put through the unit to support a given quantity of fish.

Spring water, although it may have advantages for over-wintering fish, is often too cold in summer to promote rapid growth.

The experimental lethal limit for rainbow trout is in the

region of 25−27°C. It can vary a little in either direction as a result of acclimatization.

The water on a rainbow trout farm should never exceed 22−23°C and should not exceed 21°C for more than short periods. In practice, the best possible water supply for a rainbow trout farm is one in which the temperature remains in the range of 10−15°C for as long as possible. The most vital period is from early spring to the end of summer, and the economy of a fish farm will suffer in competition with other units, if the water temperature is not within these limits at this time.

Pond culture − the Danish system

The Danish-type trout farm (Fig. 2.1) consists of earth ponds excavated on a level site. The water supply should, if possible, be by gravity, as this involves the lowest capital cost and offers the greatest measure of security. A pumped supply can be obtained from a lake or river, but this must, to some extent, be regarded as a last resort, for pumps have been known to fail and pumping costs are high, irrespective of whether diesel or electric motors are used. Further, a standby source of power is essential. For example an electrically operated pump should have a diesel motor standby, either in the

Fig. 2.1 Danish rainbow trout farm earth ponds

form of a static engine or a tractor motor. Unless really low-cost power is available, it is very doubtful if a fish farm would be an economic proposition when the water supply had to be pumped at a head greater than 2–3 m. A few fish farms are so fortunately placed as to have a substantial source of artesian water. This possibility should not be neglected.

Raceways The raceway is the original North American system designed to rear trout to restock rivers and lakes for angling. Raceways consist of brick or concrete channels, which can either be sunk in the ground, or built above ground level on a foundation of hardcore. The channels can be in continuous lengths of up to 100 m, divided by cross-walls or screens into several sections, but having a common water supply flowing through all the sections, from one end to the other of the channel (Fig. 2.2).

Raceways are relatively narrow. Widths normally range from 2 to 4 m and depths are seldom more than 1.0 m. They are sometimes constructed in double widths, with a central dividing wall running down the middle, with openings at both ends which can be closed by fish screens or stopboards. Other types include concentric, circular channels and flat spirals.

Water supply and flow Raceways depend on having a large throughput of clean water. Spring water is ideal, as fish densities are greater than in

Fig. 2.2 Raceways in S.W. France

either earth or concrete ponds. The name indicates the principle that the water must flow quickly through a raceway channel. The current which the fish have to stem all the time has some advantages when the unit produces fish for restocking as the trout will be accustomed to rapidly flowing water and will not tend to drift downstream to somebody else's water when planted out in rivers, or not to the same extent as fish reared in ponds. But the high speed of flow in a raceway is a disadvantage when trout are reared for the table market. Due to their using energy swimming against the current they will not grow so quickly, nor make such a good conversion for a given quantity of food as fish kept in more static water. This goes against the whole trend of modern commercial fish farming procedures, in which trout are being kept at the greatest possible density, with the minimum energy expenditure.

Stocking density   The density at which trout are kept in raceways must vary in relation to water temperature and available flow. A density of $4-5$ kg/m^2 and a rate of interchange of 2.5 l/min/m^2 is fairly standard, although higher densities than this are sometimes attained.

Fish tanks   Many European trout farms including those in the British Isles use circular tanks for on-growing fish to market size. The tanks are typically from 4 to 10 m in diameter and 1.6 m deep. Tanks can be moulded from glass reinforced plastic or grp and are usually prefabricated in sections. They may also be made from reinforced concrete. The tanks can come complete with bottoms or may be bought as ring walls to set on concrete bases constructed on the site.

Relatively cheap and simple tanks can be made using curved sections of galvanized corrugated iron sheet bolted together to form a circular wall. The wall may be two or more sheets high to give sufficient strength.

The concrete bases for tanks made on the site must be sloped from the circumference to a central drain. The tanks are then self-cleansing. The slope should be fairly steep without making it difficult to stand on the bottom. A difference in height of about 1 in 10 is a good compromise.

Each tank has a separate water supply controlled by a valve from the main and a screened central drain. A further development in tank culture has been the introduction of central fish grading. Each tank has a separate outlet pipe connected to a separate main which leads to a sump where the fish can be graded. The fish outlets in the tanks are closed by standpipes. When a tank is to be emptied, the intake valve is closed and the water in the tank, together with the fish, is pumped out through the grading sump. The graded fish are returned from the sump to the tanks along sections of removable pipe laid on the ground through which a flow of water can be directed to carry the fish.

Tank lay-out   Tanks are generally sunk in the ground to within about 30 cm of their full depth and firmly back-filled. The best grades of grp tank are capable of retaining their full capacity of water when erected free-standing above ground (Fig. 2.3) but unless the site is too rocky or steep they are easier to work if sunk to at least two-thirds of their height.

Construction on a level site starts with digging trenches to the full depth to take the drain pipes followed by the holes to take the tanks. The drain pipes are then laid in position in the

Fig. 2.3 Concrete tanks in Sweden

trenches. If the tanks are delivered in sections without bases, these are bolted together in the excavations. Prefabricated outlet sumps are fitted to the drain pipes. Hardcore is laid in the bottom of the excavation inside the ring wall and concrete poured in and roughly levelled. Sloping board formers are then put down and the final concrete screed is laid to slope to the central drain sump.

Fish cages   A fish cage is a netting container suspended in the water from a buoyant collar. The flotation collar is wide enough to use as a walkway (Fig. 2.4). Cages in the sea and more recently in freshwater lakes have proved to be the cheapest and probably the best way to grow rainbow trout to larger market sizes. Developments in design have allowed trout fingerlings to be stocked directly into the cages at a weight of 5 g in fresh or brackish water.

The great advantage of cage culture is that there is no risk of a failure of the water supply or lack of oxygen. There is also little danger of mechanical failure or damage provided the cages are properly anchored in a well-chosen position. The capital cost is generally a good deal less than for a farm

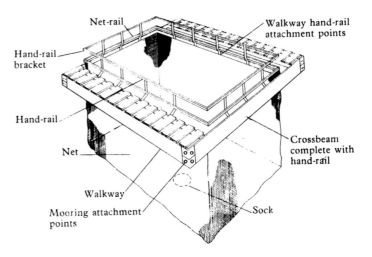

Fig. 2.4 The Kames floating fish cage

producing an equivalent weight of trout in a running water supply.

Sea cages have to be tough and well-designed to stand the wind and weather. They are more labour intensive and expensive to service than cages in freshwater lakes. The typical dimensions for a cage designed to on-grow trout is 6 m by 6 m by 4–5 m deep. The stocking density ranges from 15 to 20 kg/m^3 for fingerlings to 30–40 kg/m^3 for on-growing fish. The largest fish can be stocked at the greatest density.

Cage anchorages in the sea must be in sheltered water with a good tidal interchange. Freshwater cage sites must have sufficient depth or movement of water through the site to

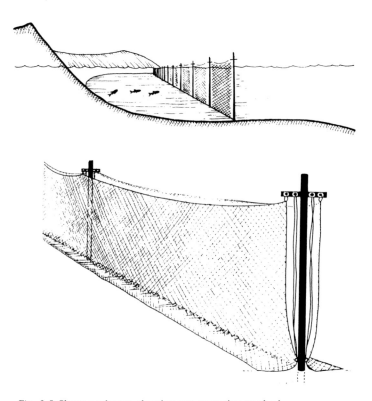

Fig. 2.5 Shore enclosure showing net mounting method

prevent the accumulation of waste matter below the cages which might cause losses of fish from self-pollution. Good sites are not easy to find in either fresh or salt water and must be carefully surveyed not only as anchorages but for access to shore-based facilities.

A disadvantage of cage culture is that flotillas of cages can be difficult and dangerous to service in bad weather. Security may be a serious problem. Large, valuable fish can be stolen from the cages. Cages in large freshwater lakes where there is little control on rod and line fishing can attract anglers to fish for the escaped trout that stay near the cage flotillas. Some of these fishermen may not stop at cutting a hole in a cage net and releasing fish to improve their chances. They usually select the cage holding the biggest and best trout.

Shore enclosures Enclosures are made of net, on a sloping shore. A double barrier made of two walls of netting is supported on poles driven into the beach at intervals of about 3 m, enclosing a semi-circular area extending along the shore below the low water mark of spring tides (Fig. 2.5). Each post has a T-shaped cross-bar at the top at right angles to the shore. The ends of the cross-bars support stretcher ropes holding up the inner and outer walls of net. Enclosures will hold 8−10 kg of rainbow trout (average weight 1−3 kg) per cubic metre of water and may be connected by tunnels of net which allow the fish to swim between one enclosure and another.

3 Brood stock and egg production

Rainbow trout remain the most domesticated member of the salmon family available for farming in temperate waters. Efforts have been made to establish desirable characteristics such as rapid growth, good food conversion, fixed age at sexual maturity and stable spawning times. Eggs can now be obtained from specialist producers that will develop into 'all female' or 'sterile' fish. The benefits of these advances to the individual trout farmer may be difficult to assess in economic terms because the situation on different farms is never the same, although the methods of fish culture may appear entirely similar.

Research is needed into what is of positive value in practical trout culture. The fact that certain strains of rainbow trout can be induced to grow very quickly is of no use if their cost of production becomes prohibitive. It is not the exercise itself which is of interest, but the ability to produce results that are demonstrably profitable. There are many factors which are of basic interest in the development of brood stock and which need to be investigated scientifically. These can be summarized as:

1 Resistance to specific diseases.
2 The ability to continue to feed effectively (to make growth) over a wide range of temperatures.
3 The ability to make useful growth on a diet low in animal protein.

The selection and keeping of brood stock is not necessarily a job for specialists and all fish farmers should know something of the basic procedures. The initial step is simple selection. The best fish are graded out, which in practical terms means those that grow to largest size in the shortest time at each stage in fresh water or in the sea. A stockman soon learns to spot the good 'doers' in the animals he tends. These are the ones to select and keep back as brood stock.

Aspects of nutrition, hygiene and disease prevention are dealt with elsewhere. The holding arrangements for brood fish are not different to those for fish intended for market but they should be kept at a reduced density.

Selection for quick growth   The factor which most commonly influences the selection of brood stock is that of quick growth, and most trout farmers tend to select and retain the quickest growing fish as brood stock from each year class (Fig. 3.1). It is, in fact, equally important, from a commercial point of view, that the selected fish should not only be quick growing, but also be good converters. Selection is not as easy as it looks and must be carefully controlled, otherwise mistakes are inevitable. If fish are merely selected because they are of larger size, this can be due to reasons other than quick growth. They can, for example, simply be a few weeks older, or have had a preferential environment and better feeding. Fry hatched from large eggs are proportionately larger than fry from small eggs, but this does not mean to say that the parent fish were necessarily quicker growing. It may merely mean that they were older, therefore produced bigger eggs.

The group of fish to be used for the selection of brood stock should come from the same fertilization, preferably made with mixed milt. The parent fish should be of the same age and as nearly the same size as possible. The fertilized eggs should be graded and those nearest to the average size

Fig. 3.1 Selected brood stock. Anaesthetized live female fish

selected. The fry should be graded early, with a first grading at about three weeks after feeding starts and then at two-weekly intervals. If possible, the selected fish should be kept in tanks of the circulating type, until the second spring. Winter grading should be carried out, if the fish are feeding and maintaining growth. Final selection should be made at or near the time the fish reach 'portion' size. If the trout farm is producing larger fish for market, either in fresh or salt water, the final selection of brood stock can be left to the third summer.

Selection for egg size

It has always been one of the accepted facts of life in rainbow trout culture that the best fry production results from large eggs. If this is true, it poses a number of problems for the practical fish farmer, as he must know how egg size varies between brood fish and to what extent this depends upon environment, feeding or genetic factors.

Selection factors

1 There can be large differences in egg size in fish of the same age group and the same parentage, but the characteristics governing the tendency to produce large eggs is not known.

2 There seems to be no relationship between the size of a female fish in the same age group and the size of the eggs that are produced. There is also no apparent way of inducing rainbow trout to produce large eggs by better feeding to promote very quick growth.

3 It seems probable that other factors besides the size of the parent fish influence the size of eggs, but at the present time no method seems to have been devised to increase egg size by altering either the diet or environment of the fish.

4 Egg size generally increases with the age of the parent fish, but this effect is most apparent in the first year after the year in which the fish first becomes sexually mature, and becomes weaker with increasing age.

5 Races of rainbow trout appear to exist with significant differences in the size of eggs which they produce but at present such specific racial characteristics have proved impossible to stabilize in brood stock.

Mixing brood stock

Each female rainbow trout will shed 1000−1400 eggs per kilogram of body weight. Normal practice is to run-off the eggs from about 10 female fish into a bowl. The eggs are then fertilized with the milt or sperm from a male fish. Fertilization takes place so rapidly that nearly all the eggs will be fertilized before there is time to take a second male. The result is that about 30 000 hatching fry will be the progeny of a single male trout.

It is possible to start the fertilization system in reverse, beginning with a number of male fish. The male fish should be dried with a cloth when they are handled. Approximately equal amounts of milt from each male fish are then run-off into a clean, dry glass container. Not a single drop of water must get into the milt mixture, otherwise the milt will be ruined. The mixed milt should be stirred with a glass rod. It can then be kept in a dark, cool place where it will remain viable for some hours. The mixture can be extracted with a pipette and used for fertilizing the eggs from selected groups of female fish.

Cryogenic storage

It is possible to store frozen fish sperm at very low temperature using liquid nitrogen. The method is basically similar to that developed for mammalian sperm. A problem has been to find an 'extender' to dilute the sperm which is compatible with the fish's seminal plasma. A satisfactory 'extender' is now available and fish sperm can be safely frozen, provided a compound such as dimethylsulphoxide (DMSO) is added to protect the spermatozoa from damage during freezing and thawing.

The commercial long-term salmonid sperm storage is likely to form a part of future stock improvement. This may be followed by the successful cryopreservation of unfertilized teleost fish eggs.

Sex reversal and induced sterility

On-growing rainbow trout to large size, particularly in floating cages in fresh water and in the sea has been inhibited to some extent by the early onset of sexual maturity, particularly in the male fish. It has proved possible to reverse the sexual differentiation of fish, which would otherwise have developed as males, by hormonal treatment given in the diet of fry during the period of initial feeding. If male gonadal tissue containing viable spermatozoa is extracted from feminized males and used to fertilize the eggs from the fully female fish, the next generation will be all female and sterile.

Disease-free rainbow trout eggs yielding all female, sterile fry for on-growing are available from commercial producers. Larger fish are in demand for processing and smoking. Male fish naturally mature early. They then lose so much in appearance and condition that they can be unmarketable in the round. The extra cost of all-female eggs is so little compared to the benefit they give it seems likely that eventually most table market trout farmers will be using them exclusively.

Buying eggs   Many rainbow trout farmers, particularly those who operate on a small scale, prefer to buy eggs from a specialist egg producer, rather than go to the trouble of establishing and maintaining their own brood stock. Eggs of early- and late-spawning parent fish are available. A rainbow trout farm, with sufficient spring water at constant temperature for incubation and the first five to eight weeks of fry feeding after the eggs have hatched, would obtain eggs of both early- and late-spawning varieties, in order to spread the hatching time, growing season and eventual attainment of market size, over as long a period as possible. It is for this reason that European rainbow trout farmers have obtained supplies of eggs from the southern hemisphere.

Freedom from   Most countries now require certification of freedom from
disease   specific fish diseases, in respect of all importations of rainbow trout eggs. Any intending importer should therefore make sure that eggs come from a registered egg producer, who can furnish a valid certificate. This is important, not only to comply with import regulations, but to protect future stocks on the importer's own and on other people's farms.

Cross   Many of the salmon-like species of fish can be successfully
breeding   crossed in the first generation, but the crosses are often sterile.

Usually this applies to both sexes, but in some cases the males or the females are infertile. Only species which are fairly closely related are likely to produce a fully fertile crossbreed.

The idea of cross breeding has diverted most trout breeders since the latter half of the nineteenth century. In more recent times some effort had been put into trying to breed fish of mixed parentage which would have specific advantages inherited from both parents, such as resistance to disease and quick growth, allied to late sexual maturity. The advent of rainbow trout farming in salt water has resulted in much effort being spent in trying to retard the onset of sexual maturity, or to breed deliberately a quick-growing, sterile cross.

Rainbow trout will cross with a number of other species, including European brown trout (*Salmo trutta*). It has been known for a long time that a cross between rainbow trout and char results in a sterile cross, which grows more quickly than either parent species. Some years ago a well-known Danish trout farmer, specializing in brood stock and egg production, established a quick-growing cross of exceptionally attractive appearance. The fish are a cross between the ordinary rainbow trout (*Salmo gairdneri*) and the Arctic char (*Salvelinus alpinus*) Fig. 3.2. This is certainly the best of trout/char varieties and should have commercial possibilities, particularly for salt water.

Working with brood stock

Rainbow trout farming is not a subject that can be learned from a book and there can be no substitute for practical experience. This gives an outline of the common day-to-day procedures on a trout farm and is intended to give some idea of what must be learned the hard way.

Stripping

It is a waste of time to try to describe in detail how to strip trout. The actual fish handling must be learned with live fish.

Eggs should be stripped from a ripe female fish, which is one in which the eggs have come loose in the body cavity and are ready to be shed. (It is easy to tell a ripe fish for anyone

Fig. 3.2 Rainbow trout × Arctic char

who has had practical experience.) The eggs are run-off into a dry plastic bowl and the fish vent should be held close to the bowl at an angle so that the eggs do not hit the side of the bowl like peas out of a pea-shooter. A little milt from a male fish is then run down into the eggs.

The proportion of males to females is a matter of opinion, of time and of the availability of ripe male fish. One male to three females is satisfactory, although a ratio of up to 15 three-year-old females has been used to 2 three-year-old males. Young males are best. Male fish can be used more than once at intervals of a day or so.

The eggs should be gently stirred after the milt has been added, left for a minute or two, and then water poured in slowly to fill the bowl. The bowl must be allowed to stand for about 10 min, when the surplus milt is washed off with successive changes of water and the eggs put aside in a bucket with about one-third eggs and two-thirds water.

Fertilization   Fertilization occurs when a male cell enters the egg through a hole in the shell, known as the micropyle, and fuses with the female cell. The eggs in the female fish's body cavity are soft, otherwise only a few would be able to fit in the small space. As soon as they are expelled water begins to pass through the shells, which are semipermeable membranes, by osmosis. The eggs then swell and the tension on the shell increases. If fertilization does not take place quickly, the spermatozoon may fail to get in through the opening. This is why the old system of 'wet' fertilization, in which the eggs from the female fish were stripped into water before the milt was added, often gave poor results. The mobility of spermatozoons is greatest in ovarian fluid; without water they will remain alive for 3.5–4.0 min whereas in water they are only active for about 0.5 min.

Eggs and sperm carefully stripped from healthy, uninjured fish, when mixed together out of contact with water, should normally result in 100% fertility. Eggs become very sensitive to disturbance for a period after water has been added and swelling starts. Losses are likely to be greatest due to movement during the first 30 min. Eggs take in water more quickly at higher temperatures. Swelling takes about 1.25 h at 6.5°C but only about 25 min at 13°C. The commonest cause of loss

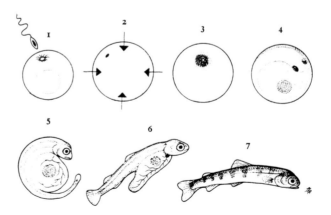

Fig. 3.3 Development: green egg to fry. (1) Fertilization; (2) Egg swells; (3) Cell division starts; (4) Eyed stage; (5) Hatching; (6) Yolk-sac alevin; (7) Feeding fry

during the sensitive period following fertilization is from washing off the surplus sperm. Losses of up to 20% can occur if the eggs are disturbed for up to 1 h after being placed in water.

When the first batch of fertilized eggs has been in the mixture of sperm and ovarian fluid for about 1 h, all the eggs that have been stripped and fertilized can be transferred to the hatchery troughs. Standing eggs must be kept cool and the bucket or egg container should be in an insulated box.

Each group of hatchery troughs should have its own litre measure. The eggs are poured directly into troughs from the measure. Eggs occupy less space before swelling and this must be taken into consideration; 1 l of swelled or 'hardened' eggs is approximately equivalent to 1.4 l of eggs before water is added.

Effect of light on eggs   Exposure to light in the blue or violet part of the spectrum can be deadly to salmonid eggs either in air or water. Eggs should be kept in the dark. When artificial light has to be provided in a hatchery this should not be from fluorescent tubes. Yellow or orange light is safe to use but ultra-violet light (UVA) can penetrate to considerable depth in water.

Disinfection   Imported eggs bought in during the eyed stage should always be disinfected. Recommended disinfectants are iodophors. These are solutions of iodine in organic solvents. The manufacturer's recommendations should be followed for trade preparations.

Buffodine is a buffered iodophor for the disinfection of eyed eggs (where the eyes of the embryo are visible through the chorion) and newly stripped, non-hardened eggs.

Anaesthetizing fish   A good, inexpensive fish anaesthetic is ethyl-aminobenzoate (Table 3.1). A commercial product is marketed under the name of Benzocaine. A concentrated stock solution can be made up by dissolving 100 g of the powder in 2.5 l of acetone. A 25 ppm working mixture is then produced by diluting 6.2 ml of the concentrated stock solution with 10 l of water.

Table 3.1 Fish anaesthetics

Chemical	Dosage (ppm)	Immobilization time (min)	Recovery time (min)	Notes on use
Carbon dioxide solid acid-bicarbonate	100−200	1−2	−	Used for killing fish
Quinaldine 2-methyl-quinoline	10−15	2−4	3−5	Non-critical Expensive
2-phenoxy-ethanol	40	2−5	5−10	More concentrated solutions up to 100 ppm may be needed for large fish in warmer water Recovery time may be longer
Ethyl-aminobenzoate 98% in acetic acid	See instructions	2−5	5−10	A good, cheap anaesthetic

Increasing the strength of the mixture will reduce the time taken to immobilize the fish.

Fish taken directly from sea cages for stripping should be washed in fresh water before being immersed in the anaesthetic solution (remember that if the fish are dripping water each time one is put into the anaesthetic solution it will soon become weak and need renewing or topping up).

Recovery should take place in a flow of clean, well-oxygenated water with the fish held upstream in the normal swimming position. It is possible to make a simple 'holder' for the fish during recovery. This consists of a rectangular trough divided into parallel channels along its length which are just wide enough to hold and support the fish. Baffles at the leading end divide the flow between the channels which are screened at the downstream end with taut netting.

Hatcheries Many different systems have been devised for incubating salmonid fish eggs. These range from troughs with longitudinal flow to batteries in which egg trays are arranged one below the other with a downward flow of water from top to bottom.

The choice of batteries for egg incubation depends upon the number of eggs which are to be incubated in the hatchery. A farm specializing in egg production or incubating more than one million eggs each winter could reasonably expect to profit by using battery incubators. On the other hand, a farm specializing in both egg and fry production would probably do better to use a standard horizontal trough and basket system (Fig. 3.4). In this system, the water flows through a trough which is 40–50 cm wide and 20 cm deep. Lengths can vary but a practical length is between 3 and 4 m. The trough carries rectangular baskets, the top edges of which rest on the sides of the trough. The bottoms of the baskets are perforated and are about 3 cm above the bottom of the trough. The bottom edge of the leading end of each basket is extended down to rest on the bottom of the trough, and there is a rectangular opening in this end, near the top, covered in perforated material. The eggs are placed in the baskets, usually about 1.5 layers deep. Water flows into the top end of the trough and the upstream end of the first basket forces the

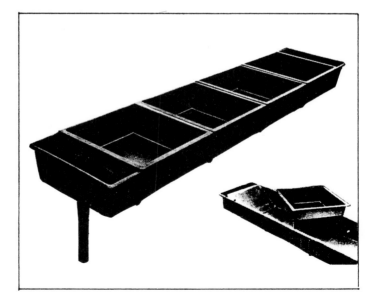

Fig. 3.4 Danish-type hatching trough and egg baskets

current under the first basket where it meets the extended edge of the downstream end. It then passes up through the perforated base, through the eggs and out through the perforated end of the basket. The current then passes down the narrow space between the first and second baskets, up through the perforated base of the second basket and so on to the end of the trough.

Each hatchery basket can incubate 12 000–16 000 eggs. The most convenient length of hatchery trough is one which will take four baskets and has a total capacity to hatch about 60 000 eggs. The quantity of freshly fertilized or 'green' eggs can safely be doubled. If the eggs are to be retained until they hatch, the density must again be reduced.

Water   The water supply to a hatchery should be completely clean and free from silt at all times. The water should be 100% saturated with oxygen and the temperature should not exceed 10°C during incubation. Rainbow trout eggs take approximately 400 day degrees to hatch and low temperatures will prolong incubation. The best water supply is from a spring or borehole where the water is at a constant 8–10°C, provided that it is fully oxygenated and is not supersaturated with air.

Flows   The flow required to incubate eggs is relatively small but depends to some extent on temperature. A flow of more than 1 l/min per 3000 eggs will seldom be needed but the installed system should provide for flows to be increased up to 1 l/min per 1500 eggs. Although a deficiency of oxygen in a hatchery, due to having too little throughput of water, may have no apparent effect on the eggs it can result in deformities or other aberrations which become apparent after hatching.

Incubation   The standard incubation procedures for salmonid eggs are quite straightforward. Removal of dead eggs, which are easily recognized because they go white (Fig. 3.5), is best carried out each day in smaller hatcheries producing fish for on-growing on the same unit. This can be done in various ways using a continuous siphon or a suction bulb. If not removed, the dead eggs will form a focus for fungus which will then attack

Fig. 3.5 The white eggs are dead and must be removed

surrounding healthy eggs. A continuous record of the dead eggs removed should be kept.

When the eggs under incubation have become eyed (the eyes of the embryo becoming visible as black spots through the egg shell or chorion) they should be 'shocked'. A simple way to do this is to lift the egg trays out of the water and leave them in the air for a few moments propped across the trough. The effect of 'shocking' is to show up infertile eggs, and any weak eggs, which will then go white and can be removed.

Specialist farmers using flask incubators, in which the eggs form a thick column in a vertical container with an upward flow of water, or hatcheries where very large quantities of eggs several layers thick are incubated in troughs, may have to use a fungicide bath to kill off fungus spores, without trying to remove the dead eggs.

Fungicides for use on fish or fish eggs

The fungicide commonly used in the past was a compound known as malachite green. This substance is now alleged to be dangerous to the people who use it over long periods (the hands of most freshwater fish farmers are generally stained green). A new fungicide has now been developed for the treatment of eggs under incubation. This is called proflavine-hemisulphate. The eggs are immersed in a solution of this compound at a strength of 1 part in 40 000 for 1 h (the arms of fish farmers using this substance are now generally stained yellow up to the elbows).

Egg counting Automatic egg counters, which also extract dead eggs, have been developed for use by specialist egg producers. The eggs are put through the machine when they have eyed-up. In one type, a disc with holes corresponding to the average size of eggs is fixed in position and then automatically loaded with eggs from a hopper as it revolves. Any dead eggs (which are opaque) intercept a beam of light as the disc, in which they are held, passes a light source. This causes a jet of water to blow the dead egg forward out of the hole into a container. As the disc revolves further round, the live eggs are also ejected into a separate container. The machine can not only differentiate between the dead and live eggs but counts them separately. The counting rate is approximately 40 000 eggs per hour. Other commercial machines can count and sort dead and live eggs at rates up to one million per hour. They are a useful tool for the large-scale operator and the cost, including purchase price, is less than one fifth that of manual picking.

Hatching and alevin development Egg shells should be removed from egg baskets after the alevins have hatched out, together with eggs which have failed to hatch. The contents of the baskets may look a mess at this time and a double-bottomed egg basket which allows the alevins to drop through a removable, perforated inner-tray is an advantage. Cleaning up after hatching is more of a problem in neutral or alkaline water. Shells and other debris dissolve and disappear fairly quickly in water of a pH below 6.5. If the hatchery water is alkaline, egg shells may have to be removed from the hatching baskets with a suction bulb, otherwise they can clog screens.

Alevins should continue to be kept in semi-darkness until they are ready to feed.

Super-
saturation   Saturation of water with oxygen means that the water has
dissolved the total volume of the gas which it is capable of
taking up at a given temperature. Water, particularly from
underground sources, can become supersaturated with gas
due to being subjected to pressures greater than atmospheric.
Water taken from below the turbines in hydro-electric stations
or from the stilling basin below the spillways in high dams can
also become supersaturated. The usual gases involved in
supersaturation are nitrogen (air) and carbon dioxide. The
effect on fish, particularly fry, is to produce what is known as
gas-bubble disease when bubbles of gas appear under the
skin. Carbon dioxide can also cause a condition known as

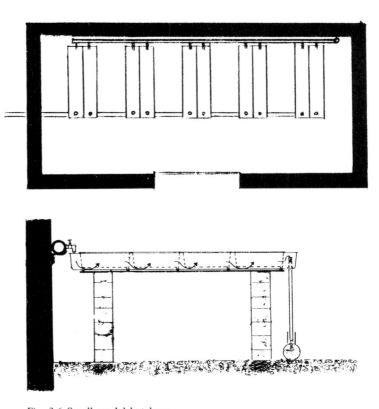

Fig. 3.6 Small model hatchery

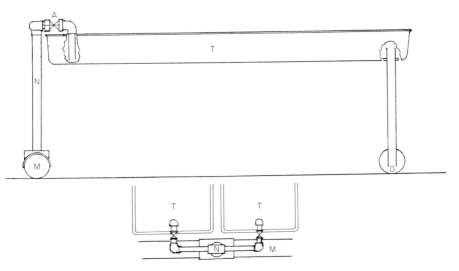

Fig. 3.7 Constructional design for hatchery trough showing water supply
and drainage. (T) Troughs; (M) Intake main; (N) Intake to troughs; (A)
Valve; (D) Main drain. Approximate dimensions: Length 2130 mm,
width 406 mm, depth 179 mm (7 ft × 16 in × 7 in). Each trough takes
four hatching baskets, each basket will hatch 6000−8000 salmon eggs
(25 000 per trough) or incubate approximately twice that number to
'eyed' stage. Flow to each trough about 10 *l* (2.2 UK or 2.64 US/Canada
gal)/min, pH 6−7.5, with dissolved oxygen never less than 70% of
saturation. Standard troughs 3650 mm (12 ft), taking seven hatching
baskets, are available. The shorter troughs are preferable if feeding is
started and the fry are fed for the first two to three weeks in the troughs

nephrocalcinosis in which calcium is deposited in the fish's
kidney.

Hatchery The model hatchery shown in Figs 3.6 and 3.7 can incubate
design and hatch 400 000 rainbow trout eggs. The troughs are 215 cm
long, 40 cm wide and 17 cm deep. Each trough has a separate
water supply taken from the 100 mm main through a 25 mm
(1 in) UPVC pipe and ball valve and is drained by a 50 mm
(2 in) vertical UPVC pipe passing through a flange at the
downstream end. The water level in the trough is adjusted by
raising or lowering the vertical pipe. The pipes from the

troughs discharge into a 150 mm (6 in) PVC drain pipe. The interior dimensions of the hatchery building are 9 m (30 ft) by 3.5 m (12 ft). The building is floored with 20 cm (8 in) of river gravel on a hardcore base. There is a wide sliding door on the long side of the building which is insulated with polystyrene sheeting. Electricity is laid on and a small industrial fan heater is mounted on one wall.

4 Fingerling production

Some trout farmers still prefer to hatch out fry and rear their own fish through to market size, but the majority buy in young fish from specialists. Farms sited where they are able to draw sufficient water of the right quality from springs or boreholes have the opportunity to produce fingerlings profitably for sale to on-growing and fattening farmers using surface water.

Fry rearing The pre-feeding fry are in the yolk-sac stage and are still feeding on a globule of yolk within the ventral surface of their body cavities. The period of absorption of the yolk sac is dependent upon temperature. The pre-feeding fry remain in the hatchery baskets until the yolk sacs are absorbed when they reach the swim-up stage and begin feeding. Some fish masters begin feeding in the baskets or in the hatchery troughs, others transfer the swim-up fry to fry tanks at the start of feeding.

It is particularly important to know the right moment at which to start feeding. This again is something which can only be learned by experience, by observing the appearance and behaviour of the fish. Feeding should begin when the temperature is over 8°C and the yolk sacs are nearly absorbed. At this stage there is a small orange bulge remaining on the belly of the fish. They will be seen to have turned into the normal dorsal−ventral swimming attitude, and no longer lie partly on their sides due to the weight of the yolk sacs.

The smallest grade of trout feed is known as crumb. The particles should be of uniform size and slow sinking. The fry can initially be hand fed if feeding is started while they are still in the hatchery troughs. Feed has to be given little and often and it is better to shift unfed fry directly into rearing tanks fitted with automatic feeders. The fish can then start and continue to be automatically fed at predetermined times with the appropriate quantity and grade of feed until they are ready to be sold or transferred for on-growing in larger tanks or ponds. Automatic feeding is simply a matter of following the directions given in the charts provided by the feed manufacturers in which the correct grade at each stage of growth is related to feeding rate, quantity and temperature.

The water supply must be checked at least twice in every 24 h and the flow adjusted accordingly. The water temperature should be monitored and it is convenient to have a thermometer mounted with a remote bulb permanently immersed in the water. An ordinary indoor−outdoor type of thermometer will function quite satisfactorily with the 'outdoor' bulb in the water. If the fry tanks are supplied with surface water, filters may have to be checked and cleaned, and the pH may have to be monitored fairly frequently, if the water is not alkaline. When acid water has to be used with a pH of less than 6.5−6.7, it may be advisable to add powder chalk to the water supply or to incorporate broken limestone in the filter system. At all times the tanks and screens must be kept clean and cleared from accumulations of surplus food and faeces.

Grading   Frequent grading was at one time regarded as essential to good fry production. The old system was to grade first at six to seven weeks and then approximately every two weeks thereafter until the fish were put out into earth ponds. The value of grading can be greatly exaggerated as the disturbance to the fish puts them off their food for a time and the loss in growth can be much more serious than the effect of having fish of different sizes in the same tank.

Production methods   Many farmers make their own fry tanks to the standard dimensions, using either reinforced concrete or a concrete skin over brick. The tanks should be constructed in a continuous run, side-by-side, sharing dividing walls. Each tank should be sloped slightly to the downstream end and must be completely drainable. A guide channel should be left in the sides and bottom to carry the screen. The water supply to the tanks should not be piped out but should be carried in an open channel along the whole length of the upper ends of the tanks (Fig. 4.1).

Rectangular tanks, 2×2 m square and $0.5-0.75$ m deep, with radius curves at the corners, have proved successful for rearing rainbow trout fry and small fingerlings. The corners must be sufficiently rounded to allow the water to circulate evenly round the tank without any dead spots. The water supply is delivered by an elbow pipe, usually directed round the side of the tank below the water surface. There is a central drain with either a flat screen leading into a sump below the tank or a vertical, cylindrical screen round a central drain pipe. The sump or drain pipe is connected under the tank to an elbow pipe at the side, which can be adjusted to control the water level in the tank. The flow to each tank is controlled by a separate screw-down valve. The tanks can either be manufactured in grp or formed on-site in reinforced concrete (Fig. 4.2).

Fig. 4.1 Fry farm — on site construction

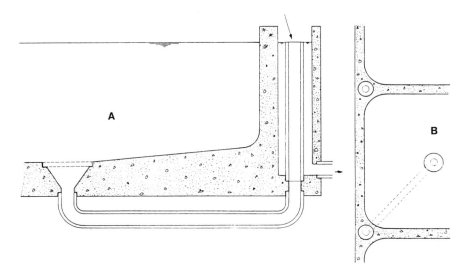

Fig. 4.2 Concrete fry tanks for on site construction. (A) Section;
(B) Plan

Round tanks may be preferred but they must be kept small
when used for fry otherwise it is difficult to keep proper
visual check on the condition of the fish. A compromise is to
use round tanks of 3−4 m diameter with a water depth of not
more than 1 m. Tanks of this size can be used to rear the fish
from fry up to large fingerlings but there is a risk of losses
which can sometimes occur in the fry stage.

The advantage of rectangular tanks for commercial small
fingerling production of fish up to about 5 g (100/lb) is that
they take up less space and can be fitted together in groups of
four which simplifies the water supply system and allows each
tank to be approached from two sides.

Water supply   The water should be as clean and free from suspended solids
as that used for incubating eggs. The temperature should be
in the range of 10−15°C and dissolved oxygen content in the
hottest weather at least 90% of saturation at 18°C (8.8 mg/*l*).

Flows and fish density   The useable area of tank base is about two thirds of the total area of a 2 × 2 m square tank (Figs 4.3−4.5) and rather more in round tanks. In the early stages of rearing, the density or total number of fry that can be held per square metre of tank area simply depends on the flow that can be given without the fish being swept away or stressed because of the energy expended in maintaining position. The temperature of spring and borehole water, or unheated surface water, should not be more than 10−12°C at swim-up when the fry are starting to feed. At this stage roughly 10 000 fry can be stocked per square metre of useable tank-base area. Each 30 000 will need a flow of about 15 *l*/min (3 gal) and water depth 15−20 cm (6−8 in). The flow into a 2 × 2 m square tank or a 3 m diameter round tank should not be more than about 20−30 *l*/min (4−6 gal).

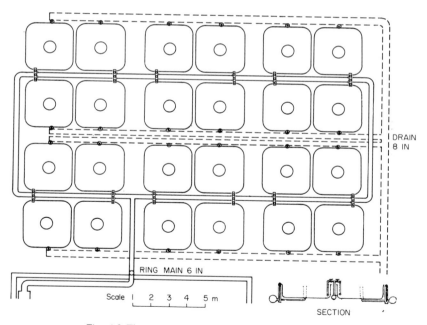

DRAIN 8 IN

RING MAIN 6 IN

Scale | 2 3 4 5 m

SECTION

Fig. 4.3 First summer rearing. Site plan of unit using manufactured grp tanks

Fig. 4.4 Fry first summer rearings tanks

By the time the fish have grown to an average weight of 1 g, the temperature of a surface water supply can be expected to be 12−14°C and the stocking density (number of fish) will have to be approximately halved. A higher density per square metre can be maintained in larger, round tanks, provided sufficient flow can be given without stressing the fish. Round tanks up to 5 m or more can be used, but the sides should not be more than about 1.0−1.1 m (3 ft−3 ft 8 in) above ground level, as this is the maximum working level at which it is possible for a person of average height to bend over the side. Deeper tanks can be used but they should be sunk into the ground.

If used to rear fingerlings to an average weight of 5 g (90 fish/lb), larger round tanks can be stocked at a density of approximately 2000 fish per square metre of useable base area, provided the water temperature does not rise above 15−16°C and the water is 95−100% saturated with oxygen. At a

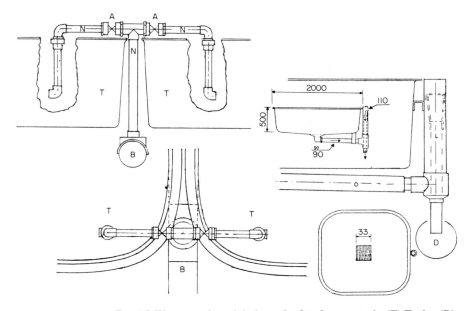

Fig. 4.5 Water supply and drainage for 2 × 2 m grp tanks (T) Tanks; (B) Intake main; (N) Intake to tanks; (A) Valves; (O) Outlets from tanks; (L) Water level control; (D) Main drain. (All measurements are in millimetres)

water temperature of 15°C and an average weight of 5 g, each kg (200 fish) of fingerlings should have an available flow of 2 *l*/min (0.44 gal).

Tank covers   Outdoor tanks should be covered by black, polypropylene netting to provide shade and protection from predators.

Indoor tanks   The cost of housing fry and fingerling tanks can be justified where wind and weather make it difficult to work out of doors. A useful compromise can be achieved with wide-span, open-sided buildings.

Sale and delivery   Fingerlings sold to on-growing farms have to be delivered without physical damage or stress. This is best carried out by a firm specializing in live fish transport. Tanks, even for comparatively short distance deliveries (50–100 miles), should be insulated. The fish are safe to travel in water at ambient temperature in cool weather provided that the water in the tanks does not warm up appreciably during the journey. The water in the tanks on the tanker-truck should have a supply of pure oxygen (spare cylinders should always be carried in case of delay) and of air from a separate petrol-driven compressor. A water pump is also essential on longer journeys where the water may have to be changed.

Market for fingerlings

The development of cage farming in lakes has created a special market for small fingerlings. These generally have an average weight of about 5 g but young fish delivered from January to March are larger and may average 10–15 g. At 5 g the fish can be contained in cage nets with a mesh of 7 mm, measured knot to knot, which is open enough to keep fairly free of algal growth.

Brood stock is selected and induced by hormone treatment to spawn at the right time. Incubation, hatching, alevin development and fry growth are governed by controlling water temperature and the duration of periods of artificial lighting. The young fish then reach the required size and can be delivered to fish farms at intervals throughout the year. Cage farmers in freshwater lakes in the British Isles and in Europe can be expected to take in stocks of fingerlings in most months except when the water may be too cold to expect any useful growth.

5 Trout culture in earth ponds

Most of the trout farms in Denmark were originally off-shoots of land farms where there was a stream that could be diverted into fish ponds. Danish trout farmers have been farming rainbow trout for almost 100 years, like any other livestock, simply in order to make a living. The water in their ponds is the enclosed space where the fish can be kept and fed to gain weight and so make a profit. Where a site can be found, on-growing rainbow trout to market size in earth ponds is still a rewarding method of commercial fish culture.

The typical layout of the Danish unit has been worked out in the light of long experience. The basic arrangement consists of a water supply channel at the highest level which feeds into ponds (Figs 5.1–5.3). The ponds then discharge into what is known as the back channel at the lowest level. The water from the back channel returns to a river, or drains away from the fish farm. Each pond has an individual water supply, which can be separately controlled, and each pond can be separately drained. The back channel has the combined flow from all the ponds.

A great many methods have been devised for controlling the inflow and outflow to ponds but the best methods are the simplest. The intake to ponds can be controlled by stop-boards, which are useful where freezing winter temperatures are expected, but it is common in modern practice to use a plastic pipe for the intake water, controlled by a valve. The intake pipe must be large enough to carry the maximum required water supply to the pond. The downstream end of the pipe must be arranged in such a way as to prevent the escape of fish, or covered by a screen.

The outlet control system known as a 'monk' (because it was invented by the monks for their fish ponds) cannot be improved upon (Fig. 5.4). A monk can be constructed in either wood, brick, concrete or plastic.

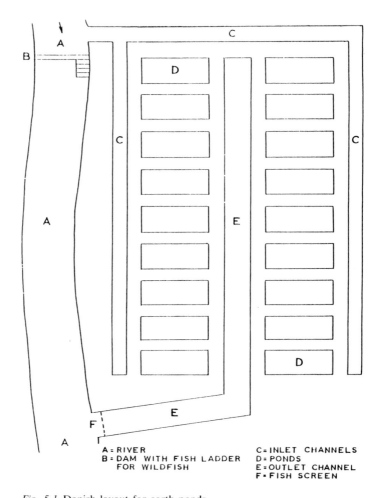

A = RIVER   C = INLET CHANNELS
B = DAM WITH FISH LADDER   D = PONDS
    FOR WILDFISH   E = OUTLET CHANNEL
                                 F = FISH SCREEN

Fig. 5.1 Danish layout for earth ponds

Pond construction   The standard Danish earth pond is 30 m in length by 10 m in width. The bottom of the pond should slope down to the outlet. The depth of the pond should be approximately 1.0 m at the upper end and 1.7 m at the bottom end. A pond of this size can be expected to carry 1.5 t of rainbow trout.

The site for earth pond excavation should be level, and the

Fig. 5.2 Earth ponds

soil reasonably impervious, with a high water table. It is possible to overcome the disadvantage of having a porous soil by puddling the bottom and sides of the pond with clay, if this is available. Some modern fish farms are experimenting with the use of reservoir liners made of reinforced plastic material.

Pond excavation on a level site requires a minimum depth of soil of approximately 2.5 m. The minimum head for standard construction, using a back channel, is 2 m. The best machine to use for pond construction is an excavator, as a caterpillar bulldozer needs too wide an area of operation. A rough test to check whether there is sufficient clay in the subsoil is whether it will form a ball in the hand. If the subsoil is reasonably stable, a batter of 1 : 1 can be left between ponds. The top of the dividing embankment must be wide enough to walk upon easily. The batter at the intake end of a pond will depend on the depth of water in the intake channel. This should be wide and shallow, in which case a steep batter can

(a)

(b)

Fig. 5.3 Earth ponds (a) Inlet channel; (b) Back channel

Fig. 5.4 'Monk' pond outlet. Sheep keep down the grass

be used for the pond face of the dividing embankment. The batter of the embankment between ponds and back channel will depend upon the consistency and stability of the subsoil. A slope of 1 : 1.5 is normally satisfactory in good soils, otherwise 1 : 2 even 1 : 3 may be needed. Surface vegetation in the top soil should be removed and not included in dam material, otherwise the embankments will become porous — this is not so important in embankments between ponds, but can certainly be a problem between ponds and back channel.

Screens Fish screens can be made in either metal, with a low corrosion factor, or, preferably, plastic. They must fit properly in their guides, and the mesh or bars of the screen should resist distortion.

It may be necessary to exclude wild fish from the intake channel, either because there are wild stocks of predatory fish in the water from which the supply is drawn, or because their entry to the ponds is undesirable for some other reason. If the water supply is drawn from a river with a natural population of salmon, there may be a legal requirement to exclude the progeny of these fish, particularly during the period of their migration to the sea as smolt.

Leaves or debris can often be a problem in the intake channel when the water supply is drawn from a river. A screening device is essential and an upward flow screen has proved most satisfactory in practice.

Screens are essential to prevent fish from escaping from each separate pond, raceway or tank. They are usually made from a corrosion-resistant metal such as aluminium, either as a wire mesh or perforated sheet. The fish screen forms an integral part of the 'monk' outlet for earth ponds. Simple vertical screens are commonly used to screen the outlets to raceways. Tubular mesh or perforated metal screens are used round the central outlet pipes in larger tanks. Small tanks used for fry or fingerlings may have a flat screen across a central sump drain.

Feeding in earth ponds   Many trout farmers prefer to feed by hand (Fig. 5.5). They can then keep a constant check on the condition of the fish. As in all animal husbandry it is essential to know how the fish are 'doing'. Experience then tells them how often to feed and

Fig. 5.5 Hand feeding

how much food to give according to prevailing weather, water temperature and the general condition of the fish.

Fish in earth ponds can be fed using 'demand' feeders from which the fish themselves initiate the distribution of their food (Fig. 5.6). Food is released from a hopper suspended over the water and a snag is that it falls in one place. The release of the food also depends on the haphazard movement of the fish and can be wasteful.

The original Danish 'gun' feeder (Fig. 5.7) fires food pellets up the length of an earth pond in a blast of compressed air. Modern versions of compressed air feeders are the best type to use on a large earth-pond farm where airlines from a central compressor can be arranged between the ponds and the separate 'guns' set for time and duration of feeds.

Grading Fish intended for grading are collected by a sweep-net into a small area at the outlet end of a pond. In Danish-type ponds

Fig. 5.6 Demand feeder

Fig. 5.7 Gun feeder

the fish can then be run out through the monk into tanks on a sorting pontoon (Fig. 5.8) floating in the back-channel. The fish are then graded into separate small tanks (Fig. 5.9) using a bar grader which has rods set closer together towards one end allowing the bigger fish to fall through first and so on until the smallest fish fall into the last tank.

Many trout farmers now use mechanical graders. The fish are lifted from the water, either by suction or a hoist, and delivered into the grader. They are then mechanically sorted into selected sizes and pumped through large-bore flexible pipes to other ponds. Central grading is generally used on larger farms. The fish are pumped or lifted with a flow of water into a container that is connected to sections of temporary, flexible piping laid on the ground or to permanently buried pipes. These take the fish on to larger tanks or raceways where they can be sorted under cover and then piped back into new ponds.

Fig. 5.8 Sorting pontoon in back channel

The same grading systems are used to select trout for sale. Fish that have reached market size and are waiting delivery are often kept in a section of the back-channel. As they are not being fed and the back-channel has the combined flow from all the ponds, they can be kept at fairly high density. Fish are often left to suffocate as they are being collected for packing. If at all possible, it is far better to kill the fish deliberately using CO_2 (carbon dioxide) discharged into the water (see Chapter 14). The fish should be netted into a small space at the outlet end of the pond or back-channel. They can then be surrounded in the net by a wall of plastic sheeting weighted along the bottom edge and the gas from a commercial cylinder released into the water.

Stocking and flows   A few old-established farms get supplies of eyed eggs, hatch their own fry and grow them to fingerlings in concrete raceways.

Fig. 5.9 Central grading tanks

Rainbow trout fry should not be reared in earth ponds due to the danger of contracting whirling disease (myxosomiasis) and should be kept under controlled conditions in concrete or grp tanks until ossification is complete and they become immune to infection. This occurs approximately 10–12 weeks after feeding starts, when the fish have reached a length of 7–8 cm. Most trout farmers buy in fingerlings at 100/lb (5–6 g) which are then grown-on or over-wintered in tanks until they are about 10/lb (45–46 g) before they are transferred to ponds.

It is generally axiomatic in fish culture that you cannot have too much water. This is particularly relevant to farming trout in earth ponds where the water temperature warms up in the summer. Standard size Danish earth ponds stocked to grow-on about 1.5 t of trout should have an available flow of approximately 300 gal/min. Less water will be needed in cold weather and when the fish are growing. The water supply should be piped from the inlet channels into the ponds and controlled by valves. The fish will do their best to jump into and swim up the intakes and the simplest way to keep them out is to put an up-turned right-angled bend at the ends of the pipes.

Aeration Reduced oxygen in the water supply to ponds during warm weather not only stresses the fish and prevents them from being fed but can also cause direct losses.

A method of aeration should be permanently installed in all the ponds and if necessary in the back-channel. Complete aeration systems are available and it is probably more satisfactory and certainly less risky to install a commercial rather than a home-made unit. Basically aeration units consist of a source of compressed air which can be piped into the water through a diffusing distributor. Sophisticated systems may continuously monitor the oxygen level or temperature of the water and turn themselves on and off automatically. Some farmers prefer to use pure oxygen either from cylinders or from an **atmos-** spheric gas extractor.

Cleaning ponds It is not possible to effect any worthwhile measure of cleansing in earth ponds containing fish. Outlet screens should be kept clean and free from accumulations of waste food and faeces. Any dead fish that can be seen in the water, on the bottom of ponds or against the outlet screens, should be removed as soon as possible. Dead fish should never be allowed to remain long enough to begin to decompose or become overgrown with fungus.

Ponds have to be emptied at fairly frequent intervals and the sludge that has collected at the bottom removed by suction into a slurry tanker. A market should, if possible, be found for the sludge which is a good fertilizer. The ponds should be left to dry out and then disinfected with a slurry of freshly slaked lime, sprayed over the bottom and sides of the dry pond. This method of disinfection offers the least risk of pollution, but the slaked lime changes very quickly into calcium carbonate, which is of no use as a disinfectant, and the slurry must be used at once. Small ponds or concrete tanks can be disinfected without being emptied but a very strong concentration of disinfectant must be used and there is a danger of pollution when the pond is drained. A concentrated solution of potassium permanganate is a cheap and effective method but the solution must remain in the pond for several hours.

Effluent from earth ponds

The discharge of the effluent from ponds takes place relatively slowly, compared to tanks where the fish are kept at much greater density with a more rapid interchange of water. The ponds themselves have a settling effect on the solids in the effluent which are the main source of pollution. This is cancelled out to some extent when pond liners are used as the solids are then discharged more rapidly (Fig. 5.10). In any event it is now usually considered essential to provide additional settlement and possibly aeration before the final effluent from a trout farm is discharged into a water course.

Fig. 5.10 Empty pond lined with plastic sheet

Fig. 5.11 'Portion' size rainbow trout

Market production

Most of the rainbow trout reared for the table market in earth ponds are marketed at what is known as 'portion' size (Fig. 5.11) when they reach an average weight of 200–230 g (7–8 oz). Growth rate depends almost entirely on local conditions. In a cooler temperate climate where ponds can freeze over in winter, small fingerlings delivered in the spring at 5 g (100/lb) and stocked in tanks, should make 45 g (10/lb) by the second half of the summer. If they are then transferred to earth ponds, the good growers should be marketable in the following spring.

A proportion of the stock may be grown-on for part of the second summer to make the larger size needed for filleting. The best fish could be graded-out at each stage and eventually supplied for fattening to large size in sea cages.

6 Raceways

The raceway was the original North American system developed for rearing trout to restock rivers and lakes for angling. The name indicates the principle that the water flows quickly through a channel. The trout were continuously exercised and were considered better able to adapt to life in the wild than fish reared in ponds.

Improved muscle tone may be an advantage in fish intended for restocking, but the energy they use swimming against the current can prevent them from making a satisfactory percentage feed conversion, compared to less environmentally active fish. Modern commercial farmers, growing trout for the table market, keep their fish at the greatest density at which they will stay healthy, with minimum energy expenditure in ponds, tanks or floating cages.

The American principle has been adopted in many other countries, but not to any extent in Europe or the British Isles. In order to be commercially viable, a raceway farm producing rainbow trout for the table market must have a very large flow of clean water with constant chemistry, clarity and temperature. These essential requirements have strictly limited the availability of potentially useful sites. In practice, the only adequate sources of water are underground rivers or large groups of springs.

Design and construction
Raceway channels are usually constructed in reinforced concrete and are sunk into the ground on a level site. Raceways can be in continuous lengths of 200–250 m, divided into sections by cross walls and fish screens, with a common water supply flowing through all the sections from one end of the channel to the other. A wide walkway between the channels allows heavy mechanical equipment to be driven up and down the length of each raceway.

Flow and fish density

A rate of water interchange of 2.5 l/min/m^2 with a stocking density of 4–5 kg/m^2 of raceway surface is fairly standard for restocking farms. The fish have to be kept at greater densities in relation to the available flow on farms producing rainbow trout for the table market. Wider and deeper raceways can hold a greater density of trout but there may be insufficient oxygen in the water to supply the fish in the lower sections of a slower-flowing raceway. This problem can be overcome by installing a permanent source of compressed air such as the large X-shaped aerators shown in Fig. 6.1 on both sides of the cross walls in the raceways.

Feeding and grading

Some of the feed given to fish in the upper sections of a raceway will be carried down to the fish in the lower sections. Each raceway has to be treated as a separate unit, stocking the top sections with fingerlings and grading the fish downstream so that they are at market size in the lowest section (Fig. 6.2). The alternative is to keep fish all of one size in the whole length of a raceway and to grade across from one raceway to the next.

Fig. 6.1 Aerators and mobile fish feeder

Fig. 6.2 Fish pump and grader

Disease risk Disease in one section of a raceway can infect fish in other sections further downstream. The fish in one section can be partially isolated for external treatment by plastic sheeting but in practice all the fish in the raceway have to be regarded as a single unit.

Investment Farming rainbow trout for the table market in raceways is not a viable option unless a site can be found that has access to the essential water supply. Given the site and the water, profitability depends on economy of scale. Only the largest farms (Fig. 6.3) can compete in price with fish from smaller units where production costs are lower. A raceway farm must make use of the most efficient mechanical aids. All grading, feeding and fish handling are carried out mechanically. The illustrations give some idea of the specialized equipment that

Fig. 6.3 Large rainbow trout raceway farm

is required. Fish feed is delivered in bulk and stored in silos. The fish are fully processed on-site and are packaged ready for retail sale before they leave the farm.

7 Tank farming

Many trout farmers in the British Isles and the rest of Europe now use circular tanks for on-growing their fish from fingerlings to market size. The tanks are usually about 1.6 m deep and from 4 to 10 m in diameter. They can be formed on-site in concrete or prefabricated in curved sections using grp (Fig. 7.1). The sections are then bolted and cemented together with mastic to form circular walls on a precast concrete base (Fig. 7.2). Glass-fibre tanks can also be bought complete with a moulded bottom, but they are rather difficult to erect on a stable foundation (Fig. 7.3).

Inexpensive tanks can be made by bolting together curved sections of galvanized, corrugated-iron sheet (Fig. 7.4). Standard width sheets can be fixed one above the other to increase the height of a tank wall. Galvanized iron tanks must be heavily and regularly coated with bitumastic paint, partly

Fig. 7.1 Grp tank wall section

Fig. 7.2 Grp tank sunk on site

Fig. 7.3 Tank showing base sloped to water and fish outlets

to avoid any risk of the fish being poisoned by zinc leaching off the surface, but mainly to prevent corrosion. Although they are relatively cheap and simple to make, corrugated-iron

Fig. 7.4 Galvanized-iron tanks

tanks can be a false economy. They can quickly corrode in acid waters in spite of being painted and have to be removed and replaced, which causes delays in planned production and upsets cash flows.

The tanks have separate water intakes taken from the main, controlled by valves. Figure 7.5 shows a simple and effective valve that can be made in plastic piping from a T-junction fitted over a slightly smaller-bore pipe. The inner pipe has a hole cut in one side and can be turned to open or close the outlet and control the flow (Fig. 7.6).

Tank farms are usually designed for central grading. Each tank has an additional outlet pipe, separate to the water drain (Fig. 7.7). The extra pipe leads to a sump where the fish can be graded. When the fish in a tank are to be sorted, the water intake valve is closed. The fish outlet is opened and the water from the tank carries the fish down the pipe into the grading

Fig. 7.5 Low cost intake valve

Fig. 7.6 Intakes to grp tanks

sump. The fish can be stopped by a screen and hand-sorted or pumped out with the water into a mechanical grader. The sorted and graded fish can then be returned in a flow of water

Fig. 7.7 Tank with water and fish outlets closed with stand-pipes

to appropriate tanks through removable sections of flexible pipe laid on the ground.

Water supply A tank farm should take water by gravity from a clean river. Pumping is expensive. Electric pumps can fail and a diesel stand-by generator with automatic switching is essential.

Fish density and flow Ordinary tanks with a water supply at temperatures 15−12°C could be stocked at densities of 25−35 kg fish/m of water. If higher temperatures can be expected, initial stocking may have to be at reduced density or a proportion of the fish moved into spare tanks.

In water at 15°C an available flow of approximately 125 *l*/min will be needed for each 100 kg of fish. A flow of 200 *l*/min should be available if the water temperature is likely to reach 20°C.

Tank farm construction The first step is to survey the site accurately. If the fish farmer cannot use a level, the job must be done professionally. The intake pipeline, tank layout and outfall are then drawn out to scale in plan and section showing the levels.

The holes for the tanks and tracks for the pipes are excavated with a digger according to the plan. The pipelines for water and fish extraction are laid down. Complicated concrete work in on-growing tanks can be avoided by using exterior fish screens. The outlet flow pipe is extended up outside the tank into a sump beside the tank. The water flowing from the sump back into the outlet main passes through a screen.

The tanks are then assembled on a level base in the excavated holes. The sections are bolted together with a sealing compound laid along the joints. Stand pipes, long enough to come above water level, are push-fitted into the open ends of the right-angle bends in the water supply and fish pipes, in the middle of each tank base area.

A layer of hard core is levelled into the tank base areas and concrete is poured to provide an overall depth of 0.15 m (6 in). Further concrete is added to provide a sloped base of about 1 in 7 from the centre to the periphery. Wood formwork is used to produce the required batter. The tanks can then be backfilled leaving about 0.6 m (2 ft) clear above the ground.

Low-head recycling
The most economic source of water supply to a tank farm is by gravity from a clean river, but it may be impossible to return the effluent without damage to the environment. Treatment is expensive and a compromise is to use less water and so reduce the quantity that has to be treated before being discharged. This is possible by partial recycling (Fig. 7.8).

A trout farm based on the design shown in Fig. 7.8 works with a gravity water supply of about 0.45 m (18 in) between the intake and the first rank of tanks (Fig. 7.9). The effluent water from the first rank of tanks flows into a 'separator' (see Chapter 13) which removes most of the solids. Oxygen is then injected into the water which goes to supply the second rank of tanks. The head between the two ranks is about 0.45 m (18 in).

The final effluent from the farm goes through another 'separator' and may also require to be re-oxygenated before being

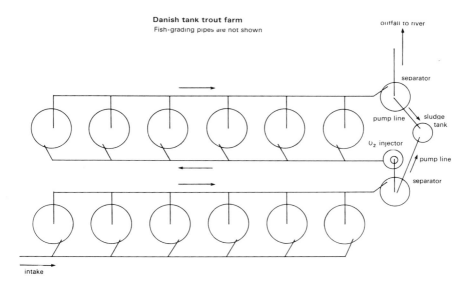

Fig. 7.8 Layout for low-head recycled water supply

Fig. 7.9 Low-head recycling on a Danish trout farm

discharged. This system will on-grow about twice the weight of trout that can be grown by single use of the same quantity of water. The reduced abstraction combined with effective effluent treatment is more likely to meet with official approval.

Solar heating   Salmonid farming has generally developed in comparatively cold climates, where the water is cool enough for the fish in summer but too cold for them to feed enthusiastically in winter. The optimum temperature for growth only lasts for short periods. It is possible to provide artificial heat, but where the water is flowing through and changing continuously it is too expensive, except perhaps for hatcheries and fry rearing units. It has been used for closed-circuit type intensive recycling culture but even then it adds considerably to production costs. An interesting possible development is to use solar heating. This has only been tried experimentally, using an array of panels erected beside raceways where the water was recycled. It could be a valuable means of generating sufficient heat to maintain optimum water temperatures for growth in intensive or semi-intensive production.

8 Cages in fresh water

Cages in freshwater lakes have proved to be the simplest and most profitable method of farming rainbow trout for the table market, but potential sites are very few and far between. The lakes must be large and very deep so that the water stays cool in summer and remains comparatively warm during the winter. The fish held in floating cages will then maintain more or less continuous growth. The chemistry of these lakes in areas of igneous or hard metamorphic rock promotes a negligible algal bloom and the cage nets remain clean. The water is free from suspended solids and there is no risk of loss due to pollution. The pH is constant and the fish are generally healthy in a stable environment.

Sites In the British Isles, the very large lochs (lakes) in Scotland have provided good conditions for freshwater cage culture. Some Scottish freshwater cage farms are now producing 200−300 t of trout a year, without damage to the environment. The lochs that have proved profitable for cage farming are in geological faults. They are very deep, going down to depths of 300 m or more which is well below the present sea level. The great body of water stabilizes the temperature which does not fall below about 5−6°C in winter or rise above 15−16°C in summer.

Anchorages The anchorage should be sheltered by a point of land or in a bay from the prevailing wind and waves travelling for more than 1 km. The exact position of the lines of cages must be carefully surveyed. The bed profile of the deep Scottish lochs that have proved suitable for cage culture extends out from the shore in a comparatively shallow shelf for about 50−200 m. It then descends very steeply down an almost vertical slope to great depth. Flotillas of cages can be anchored in lines parallel to the shore at the verge of the steep decline and waste food and faeces, collected below the cages, descends through a very long water column that acts as a biological filter (Fig. 8.1).

Collars and cages

The fish are held in an enclosure of net supported by a flotation collar. Simple, rectangular collars are usually used for cages in fresh water. When the cages are joined together, the sides of the collars form walkways which are wide enough to take the gear needed for grading and other servicing.

A fish cage is a box of net with rectangular sides and a base that may be flat or tapered to a funnel that can be opened and closed to allow debris or dead fish to be removed. Mounting ropes are sewn in as the cage is made. The net material is braided nylon or terylene twine, either knitted into a knotless mesh or knotted in the ordinary way. A knotless mesh is best as it offers a smoother surface to the fish.

Fig. 8.1 Freshwater cage site

A useful size of freshwater fish cage is approximately 6 m × 6 m × 4.5 m deep. Cages of this size can be used for fingerlings of 5 g (100/lb) up to market fish. The mesh for fingerlings is 5−7 mm, measured from knot to knot along one side. The mesh for on-growing to portion size is 12 mm, and 20−25 mm for large fish.

Servicing The cages are most easily serviced using a large, stable raft which is simply a platform supported on oil drums. The raft should have a protecting rail round one end and a mounting for an outboard motor. Service rafts must be large enough to carry substantial amounts of feed and equipment.

Shore bases The base should if possible be close to a public road. A large fixed or floating landing stage is needed that can be approached by a roadway wide enough to take a fish transporter (Fig. 8.2). A building will be needed for icing and packing fish with room for an ice-maker and a large store for nets and gear. There should be a separate food store or silo. A diesel

Fig. 8.2 Landing stage for securing freshwater cages

generator and fuel tank will be needed if there is no source of
mains power.

Fish feeding   Some form of automatic feeding is necessary in all cage culture
if the cages cannot be reached by a walkway extending directly
to the shore. The best type of feeder has separate, built-in
controls for dispensing food (Fig. 8.3). Automatic feeders of
this kind are usually governed by light-sensors which turn on
and off the power source according to the amount of light, and
have timers that release the feed at pre-determined intervals.
Large food hoppers should always be used to cut down the
frequency of loading.

Grading and   Ropes sewn in below the cages are pulled tight to raise the
killing   nets and divide them into two or more shallow sections. Fish
can then be graded using dip-nets and ordinary bar-type
graders. A lightweight, automatic grader can be carried on

Fig. 8.3 Automatic feeders on freshwater cages

the service raft, with a water pump driven by a small petrol engine, which returns the fish to cages through flexible pipes. Fish for market can be loaded into a container on the service raft through the grading pipes. It is preferable to kill the fish while still in the cage by discharging carbon dioxide into the water. The cage has to be surrounded by a wall of plastic sheeting and the net is raised to confine the fish in a smaller space. The CO_2 gas is introduced from a cylinder, through a diffuser on the end of a pipe inserted below the net, inside the plastic sheeting so that the gas bubbles up through the fish.

FRESHWATER CAGE FARM PRODUCTION

Annual capacity
Cage type: Kames 3 t cages − 100 m^3 capacity
Cages: 42 operational plus three for grading and culling
Finish fish-weight/density: 25 kg/m^3 (in cold water during winter the Kames 3 t cages will carry 35 kg/m^3)
Annual production: 200−250 t of rainbow trout
Average weights of market fish: 340−400 g (12−14 oz) (fish mainly for central processing into pre-packaged fillets)

Fingerlings
Fish at 100/lb (5 g) are stocked at 100000 per 3 t cage, then split at 0.5 oz (10−15 g) and stocking density reduced to 50000 per 3 t cage, weight/density 8−10 kg/m^3

Cage mesh
Cage mesh with fish at 100/lb (5 g) is 7 mm measured knot to knot. The fish when split at 0.5 oz (10−15 g) are transferred to 12 mm mesh cages and remain there for on-growing to slaughter.

Fingerling deliveries
100000 fingerlings, 100/lb (5 g) are delivered from specialist suppliers in each month of the year from March to October.

Summer growth
Fingerlings delivered in March/April will make 350−400 g (12−14 oz) by November−December of the same year (11 months on-growing).

Winter growth   Fingerlings delivered in October will make 350–400 g (12–14 oz) by September of the following year (13 months on-growing).

Feeding   With careful attention to weather conditions the fish will continue to take food and make profitable growth at surface-water temperatures down to 3°C.

Stocking   'All female' fingerlings are only delivered for the last two batches (September–October) on each complete production cycle. This is to avoid precocious males in the following year. Otherwise fish of random sex are stocked, as males make better early growth.

Water temperature and oxygen   Lack of oxygen in the warm surface water of a lake, round a flotilla of fish cages during hot weather in summer, has to be compensated for by the use of separate, motor-driven aerators in the cages. This is expensive in capital and running costs. Ice formation in winter can only be prevented by finding a site for the cage anchorage where springs in the lake, or the incoming flow from a warmer spring-fed river, warm the surface. It is possible to overcome either of these problems by pumping up deep water and directing this to flow through the cages. Water from sufficient depth has a stable temperature being cooler and well oxygenated in summer and warm enough not to freeze in winter.

The system is quite straightforward and the detailed method adopted must be planned to suit the site. A large capacity engine-driven pump is mounted on a raft anchored over deep water close to the cage flotilla which seasonal sampling has shown to be clean, and have the required stable temperature. The pumped-up water is then discharged through separate lightweight metal, perforated pipes mounted vertically below the walkway at the windward end of the cage flotilla. The perforations on the discharge pipes are designed to aim a flow through the cages.

9 Rainbow trout farming in salt water

The main reason for sea farming is the unrestricted water space. In the colder, northern countries there is the added advantage that the sea is usually relatively warm in winter. The fish will continue feeding through the winter months and make much more rapid growth than in fresh water. European coastal waters warmed by the Gulf Stream seldom fall below 5°C through the top 5 m of depth, and in some sheltered inlets and sea lochs the water temperature can remain in the region of 9−10°C for most of the winter.

Rainbow trout of the ordinary commercial strains do not tolerate life in salt water of full oceanic salinity until they are well past the fingerling stage. Sea farms should therefore be producers of large rainbow trout. Early sexual maturity, particularly in male fish, caused problems but these have now been overcome by the use of all-female stock.

Some sea farmers now consider that farming big rainbow trout in salt water can be a better economic proposition than farming salmon. Trout grow faster than salmon, cost less to rear and the husbandry involved is relatively simple.

Saltwater tolerance The body fluids of trout and salmon have a salt concentration approximately equivalent to one part sea water to two parts fresh water. While the fish remain in a freshwater environment, water diffuses into their tissues and other parts of the body surface where a semipermeable membrane comes in contact with the water. The fish discharge the surplus water as urine. The situation is reversed when the fish are in the sea or in water more saline than their own body fluids. They are then continuously concentrating a solution of salt in their bodies. The salmonid fish species excrete the extra salt through special cells in the gills. In the migratory 'steelhead' race of sea-going rainbow trout the salt-excreting cells increase in number when the fish undergo the change into smolt. This helps them adapt to life in salt water. The number of salt-excreting cells is very much less in the races which do not migrate to the sea and spend their whole life cycle in fresh water. The number of salt-excreting cells can be artificially increased by feeding a high-salt diet to the fish while they are still in fresh water.

Acclimatization to sea water

If at all possible, the transfer of rainbow trout from fresh water to sea water (30−35‰ of salts) should be carried out following a period of acclimatization in water of gradually increasing salinity (Fig. 9.1).

Some sea farms introduce the fish to mildly saline water early in the first summer. The shore tanks have a dual fresh- and saltwater supply. Small rainbow trout fingerlings will adapt to a mixture of two parts fresh water to one part of fully saline sea water. They will then tolerate a mixture of one part fresh water to one part sea water by the end of the first summer and water of full oceanic salinity in the spring of the following year.

Tolerance to full sea water of unacclimatized fish depends to a great extent on their size. Some rainbow trout of ordinary commercial genetic background will tolerate full sea water by the time they reach a length of 20 cm and a weight of 100 g. When the fish have reached a weight of 150−200 g the percentage of casualties on transfer to salt water is usually quite small, but some casualties are inevitable unless the fish are acclimatized or have been grown from the fry stage in a mixed water supply of gradually increasing salinity.

Shore farms

The first Norwegian saltwater farms carried out the complete rearing of rainbow trout from fry to market size in tanks on the shore. An increasing proportion of sea water is pumped into the tanks at each stage of growth. The freshwater supply to the tanks holding the smaller fish is by gravity. Figure 9.2 shows a simplified layout of one of the original units.

The main drawback to the shore-based system is the cost of pumping large volumes of sea water. This is not so serious in Norway where there is only a small tide of less than 1 m in most of the fjords and electricity is relatively cheap. In spite

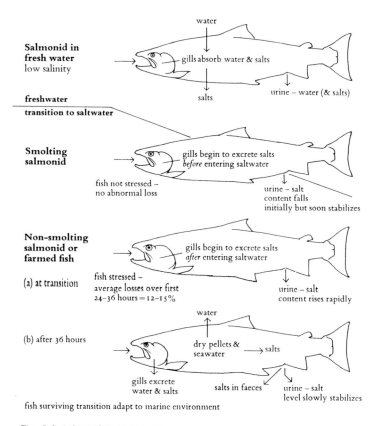

Salmonid in
fresh water
low salinity

water

gills absorb water & salts

salts

urine – water (& salts)

freshwater
transition to saltwater

Smolting
salmonid

gills begin to excrete salts
before entering saltwater

fish not stressed –
no abnormal loss

urine – salt
content falls
initially but soon stabilizes

Non-smolting
salmonid or
farmed fish

(a) at transition

gills begin to excrete salts
after entering saltwater

fish stressed –
average losses over first
24–36 hours = 12–15%

urine – salt
content rises rapidly

(b) after 36 hours

water

dry pellets &
seawater   → salts

gills excrete
water & salts

salts in faeces

urine – salt
level slowly stabilizes

fish surviving transition adapt to marine environment

Fig. 9.1 Adaptation to seawater

of the high cost, this method of farming is still being adopted
in other countries, mainly because servicing is easier on the
land.

There are other drawbacks to shore-based saltwater farming.
Pumps and motors have to be protected from corrosion. Pipe-
work becomes fouled with marine growth and duplicate lines
may be needed to allow time for cleaning. The tidal flow must
be sufficient to shift the effluent, and the outlet must be far
enough away from the intake to prevent recirculation. Floating
intakes suspended below a raft may have to be used to prevent

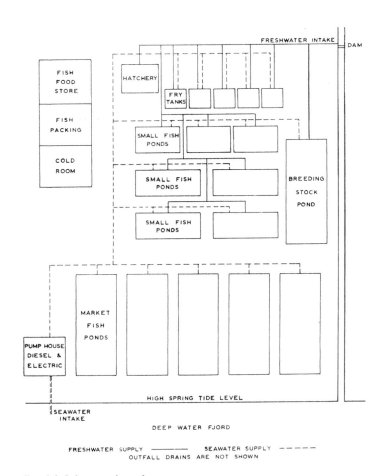

Fig. 9.2 Saltwater shore farm

blockage by debris. Suitable sites for this kind of operation are comparatively few and far between but, in spite of the drawbacks and economic limitations, the method still attracts some newcomers to the industry.

Tidal enclosures   This system is only practicable in a limited number of places where there is a small tide but a strong tidal flow. The tidal flow has to be sufficient to change completely the water in an enclosure during the course of each ebb and flow. The enclosures are made of a double wall of net supported on poles enclosing a semicircular area of water extending seawards from the low water mark of ordinary spring tides. The net walls are held down by stones or by lead-lines sewn into the bottom of the nets. The double wall allows the net to be changed when it becomes fouled by marine growth. A well-sited enclosure with a good tidal interchange will hold 8−10 kg of rainbow trout per cubic metre of water, averaged over a tide cycle, while the fish are grown-on to a weight of 3−4 kg.

Sea cages Cages for on-growing in the sea or in brackish water are constructed in the same way and serve a similar function to cages in freshwater lakes. They consist of a flotation collar which can be rectangular, polygonal or circular and may incorporate a walkway (Fig. 9.3). This supports a deep cup of netting to hold the fish. There may be a top net to prevent the fish jumping out and to keep off predatory birds (Fig. 9.4). The flotation collars for sea cages are generally stronger than those used in fresh water as they have to stand up to greater stresses in bad weather and they are usually larger. Big sea cages can have a capacity of more than 700 m^3. Very large cages are risky on smaller sea farms.

Anchorages and shore bases Hydrographic factors govern the selection of a site. The 'fetch' or distance of open water over which the wind can build up a wave should not be more than about 1000 m. The depth should be at least 15 m at low water and the tidal flow not exceed 0.5 m/s on spring tides, otherwise the net cages will be pulled out of shape. A scoured sea bed may indicate that the tidal flow is too strong, a muddy bottom show that it is too weak. The full sea survey must investigate currents at and below the surface, and the range of salinity and temperature, not just over a short period but over the year.

The principal capital cost of a sea farm is the development of the land base. This will be a dead loss if there is no security of tenure over the sea-cage anchorage. The lease of the sea bed must be legally sound and of at least 20 years duration

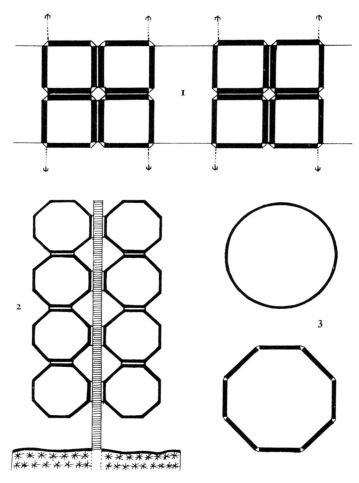

Fig. 9.3 Sea cages (1) Small cages joined in flotillas; (2) Walkway cages; (3) Large round and polygonal cages

with a predetermined rent agreement. The shore base must be accessible by land and water. The main consideration should be how and where the fish are to be sold.

Fig. 9.4 Sea cages with automatic feeders and top net

Cages on a shore walkway
A long, floating deck extends from the shore with fish cages anchored along both sides. The advantages of this system are ease of servicing and good security. The flotation for the cages moored alongside the stable working platform can be lighter than those needed for an off-shore anchorage. Cheap rigs made of galvanized pipe-scaffolding supported by tar barrels have made a lot of money for some ingenious farmers.

Simple collars can be made up from large-bore, foam-filled plastic piping which is virtually indestructible. Fish feeding is easy and compressed-air feeders can be used, mounted on the floating dock, with flexible airlines to a compressor on the shore. Grading and fish handling is straightforward and the cages can be serviced without fear of wind and weather.

The disadvantage of this type of unit is the risk of self-pollution of the site. The water must be deep enough and the tidal flow sufficient to clear the waste products from below the cages. Suitably deep and sheltered sites are usually only to be found along a mountainous, fjord coast protected by off-shore islands.

Floating 'islands' The method is intended for fairly large sea farms with a production target of 200–250 t of trout a year. The whole fish servicing and security system is built on a very large raft or floating 'island' (Fig. 9.5). The raft has a central working area carrying fish food, fuel and equipment stores, compressor and generator, food preparation room and guard room. The fish cages are moored alongside the 'arms' and there is a dock for a work boat at the end of one arm.

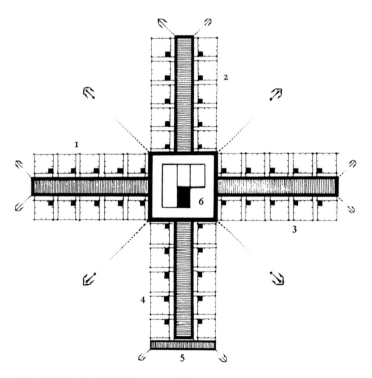

Fig. 9.5 Floating 'island' sea-cage farm

Fish husbandry

The technique for feeding and grading fish in floating cages anchored off-shore in flotillas is straightforward provided the flotation collars themselves or a central raft provide a stable working platform. Automatic hopper-feeders can be used and the fish are graded using ropes sewn into the bottom of the nets to lift and divide the cages. The fish can be grown-on at densities of $30-40$ kg/m^3.

Maintenance

Servicing and general maintenance of cages in deep water requires some underwater work. This is best carried out by staff trained in the use of SCUBA gear. Fouling can be a serious problem. Cage nets become clogged with marine growths to such an extent that water fails to circulate freely through the mesh. A number of different antifouling treatments are available for use on the site and new nets can be factory treated before delivery. Badly fouled nets must be changed and cleaned on shore. Immersion in fresh water is the best way to kill the marine animals. The nets can then be left to dry suspended in the air on ropes between poles on a 'drying green'. Nets can be obtained with a strand of copper wire in the twine which may inhibit some marine growths.

Sea fattening

The difficulties of transferring small rainbow trout to salt water has brought about a new approach to sea farming. The concept behind the on-growing of fish of freshwater stock in sea cages is now more properly regarded as straightforward fattening. The trout are comparable to 'store' lambs or other livestock that are sold when they reach a certain size by one farmer to another who has the means of fattening them for market. The 'store' rainbow trout are large enough to tolerate marine salinity, and at the optimum size for rapid growth when they are bought in and transferred to sea cages. The profit margin for the sea farmer is the weight gain of the fish, less the cost of fish feed and husbandry.

Weight at transfer to sea cages

The trout are usually moved into the sea at $700-800$ g. Trout whose early growth has been accelerated in warmed fresh water, have been found to grow more quickly in the sea than older fish that have taken longer to reach the same size.

Weight gain in the sea

Trout weighing 700—900 g when transferred to sea cages in April can reach 3—3.5 kg by December of the same year when they are 22 months from swim-up (starting to feed in fresh water). The problem of advancing sexual maturity can be overcome by using all female stock rather than sterile or triploid fish. The daily percentage weight gain for all female fish is about four times greater (faster) than for triploids.

Osmotic stress

Even though domesticated rainbow trout of mixed genetic descent tolerate sea water fairly well at the weight they are transferred for fattening in sea cages, any losses are expensive. It is not so much failure to osmoregulate that kills the larger fish, but diseases such as vibriosis and furunculosis which take advantage of osmotic stress. The amino acid thyroxin plays an important part in the physiological changes of smoltifying salmonids. Osmotic stress, and consequently losses from stress-induced diseases, can be reduced from 6—7% in untreated fish down to 2—3% by feeding 0.5 mg/kg of thyroxin for 2 months before transfer to salt water. This only adds about 20% to the cost of the feed over the eight weeks, which is of no consequence compared to the safety margin it can provide.

Steelhead or sea-going rainbow trout

Sea farming of rainbow trout is wide open for development. The fish can be kept at a much greater density than in fresh water and grown to a size which is comparable to that of small salmon or grilse. It is obvious that the best race of trout to use for marine aquaculture is one that is naturally adapted to salt water.

It is known that ordinary domestic rainbow trout, of the varieties reared in fresh water for the table market, will tolerate salt water when they reach a certain size. A greater certainty of successful transfer is likely if the stock is derived from parent fish that are not merely tolerant of salt water, but naturally designed to spend their main growing period in a marine environment. Steelhead (Fig. 9.6) are the obvious choice, and the establishment of a racially pure brood stock, derived from wild parent fish of the sea-going form, either winter or spring spawning, would be of great advantage to the marine rainbow trout farmer.

Fig. 9.6 Steelhead sea-going rainbow trout

Rearing   Wild steelhead trout have been reared to the smolt, or sea-migration stage in captivity, using ordinary crumb and pellet dry feeds. The majority (80% approx.) of young, wild steelhead become smolt and migrate to sea in the spring or early summer of the second year (scale age 2). The mean length at migration, regardless of age, is approximately 16 cm, and the mean weight at migration, also regardless of age, is 43.2 g. It is the length of the smolt of the sea-going species of salmonid fish that governs their age and time of first migration to the sea. Steelhead, reared in captivity, which can be brought up to a mean length of 16 cm in one feeding season, will be ready to tolerate full sea water (34−35‰) one year earlier than the majority of wild, steelhead smolt.

Growth rates   See Table 9.1.

Table 9.1 Average growth rate, measured from samples of wild steelhead trout, returning from the sea to river

River	Race	Years of sea feeding	Average weight
Alsea (Oregon)	Winter run	1	1.13 kg (2.5 lb)
Alsea (Oregon)	Winter run	2	4.08 kg (9 lb)
Babine (BC)	Summer run	1+	2.3 kg (5 lb)
Babine (BC)	Summer run	2+	5.2 kg (11.5 lb)
Babine (BC)	Summer run	3+	9.2 kg (20.25 lb)

Steelhead culture

The brood stock generally used for hatcheries in Oregon and Washington are winter steelhead. Eggs stripped in early January will be 'eyed' by the beginning of February and hatch in March. The monthly hatchery-water temperatures are approximately January, 7°C; February, 8.5°C; March, 8.5°C; April, 9.5°C. The average water temperature from 'green' (freshly fertilized) egg stage to 'swim-up' (starting to feed) is 8.4°C.

Fish hatched in March that start feeding in April can reach a length of 16.0 cm by October of the same year. If the seawater temperatures are high enough for the fish to continue feeding through the winter months, they should make 350–400 g in the following spring and 2.5 kg by the end of the summer. The overall rate of growth depends almost entirely on an early 'swim-up' following accelerated incubation in comparatively warm hatchery water.

Recirculated water

The initial development of fish of the salmon family can be accelerated by using naturally warmed borehole water, but this may not be sufficient for commercial fry production unless it is recirculated. The design of a foolproof, recycling system, with the essential high level of serviceability, requires special technical knowledge and mechanical expertise. It is probably safer to buy a ready-made unit which is guaranteed by a reputable manufacturer.

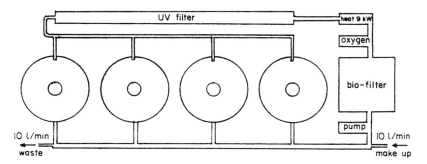

Fig. 9.7 Diagram of recirculation system

The basic design of most recycling units consists of a fish tank from which the water either flows under gravity or is pumped through a filter to remove suspended matter (Fig. 9.7). The water then passes through a biological filter, often a vertical silo packed with plastic pipes which become coated with the essential purifying bacteria. Oxygen or air is injected into the water which may then be heated to raise the temperature to the optimum for the fish and treated under UV light to kill off disease organisms. A make-up of fresh water consisting of about 10% of the flow in circulation is introduced in the course of a cycle.

Rainbow trout of the races usually farmed for the table market in the British Isles and in continental Europe do not fully tolerate transfer to seawater of oceanic salinity. On the other hand they do particularly well and grow very fast in brackish water with a salinity between 15 and 20‰ (1.5 and 2% salts in solution). It is important that the salinity should not vary greatly due to tidal interchange or inflows of freshwater. These conditions have, to some extent, limited the possibilities for large-scale saltwater rainbow trout culture in the British Isles. There are very few low-lying coastal areas where there is a large body of brackish water with a sufficiently constant level of salinity in the optimum range.

Some of the larger sea lochs on the west coast of Scotland have potential sites, but very few of these have an area where the salinity remains relatively stable for any length of time.

There are some coastal waters in Ireland which could prove suitable, but interest has been concentrated on farming Atlantic salmon and little attention given to investigating sites for fattening rainbow trout in saltwater. So far development has been mainly concentrated in Scandinavia, particularly in Denmark, where sea-farming rainbow trout is now an increasingly important branch of the industry.

10 Trout farm design, equipment and operation

The development of trout farming over the years has inevitably produced many widely differing holding systems for trout and an equally wide variety of equipment intended to speed up or simplify husbandry procedures. The methods for on-growing can be divided into non-intensive, semi-intensive and intensive. Earth ponds, big tanks and large floating cages are basically designed to work most efficiently with low-density stocking. Medium size tanks and cages can be stocked more densely and are more economical in space but they need more water and a quicker interchange. Complicated and expensive recycling systems are only profitable if they can be stocked at the highest densities. They need a very high standard of husbandry and hygiene as the fish are always at risk from mechanical failure or human error.

Cage farming is the lowest cost method of producing trout. The design of a cage farm is a relatively simple matter of selecting the most suitable flotation collars and working platforms. The difficulty is in finding a site. These are very few and far between, and most aspiring trout farmers must build their farms on a clean river from which they can draw a gravity water supply. The problem is how to return the water from the fish farm without damage to the environment. The final effluent may have to be treated, and to do this cost effectively it may be necessary to recycle a proportion of the flow (see Chapter 7).

The simplest systems of rearing trout for the table market may no longer be the most cost effective. Over the last 20 years sophisticated mechanical aids have become available which can now be used to improve conditions for the fish and to keep down labour costs on all but the smallest farms. Purpose-made equipment can be obtained to perform most of the day-to-day operations on a trout farm that used to be carried out by hand.

Sites A site survey and full investigation of local conditions must be the first step taken in planning a trout farm. Potential sites for sea cages need to be explored initially to discover exposure and shelter, depth and bed material, tidal flow and currents, average monthly salinities and water temperatures, access and the location of any other fin-fish or shellfish farms in the vicinity. The essential features of sites for freshwater cage farms are outlined in Chapter 8. The investigation for trout farms using running fresh water must cover the geography of the site and the average monthly flow and temperature of the available water supply. The water should be sampled for analysis in spate and in drought flows.

Viability The initial stage of planning must involve a careful and uncompromising feasibility study in order to decide if the project is viable. Start with where, how and for how much the fish can be sold, then work back to see how much they will cost to produce, using the most economical method applicable to the site.

Operational equipment Methods of operation must be considered at the design stage as some essential equipment will need to be incorporated in construction.

Aeration A means of providing additional air or oxygen to the water supply may have to be included in the lay-out of ponds, raceways or fish tanks. Aeration or oxygenation may be required to improve the quality of the effluent water before it is discharged into a water course.

The water from deep boreholes or from below generating stations can be supersaturated with air (nitrogen gas) in solution. This will cause gas-bubble disease in hatchery fry and has to be removed by cascading or blowing compressed air through the water.

Acidity   Acid precipitation or natural acidity can cause either episodic or continuous lowering of the pH to a level that is lethal to fry and damaging to larger, on-growing fish. A simple method of preventing losses in earth ponds is to install a 'chalking' mill over the intake channel. A water wheel drives a mechanism that releases powdered chalk into the water. A more sophisticated means of maintaining or restoring alkalinity may be needed for the water supply to raceways or fish tanks.

Screens   A screen to exclude debris will be needed on the intake channel to earth pond farms. The best type is the upward flow screen shown in Fig. 10.1 which can be cleaned by backwashing through a by-pass sluice. Fish screens will be needed at each pond outlet. A prefabricated screen to fit on a 'monk' is shown in Fig. 10.2. The back channel will also require fish-screening, preferably with a self-cleansing drum-screen (Fig. 10.3).

Filters   Filtration for the whole water supply to an on-growing farm cannot be justified economically. Screens may be essential for hatcheries using surface water. The most satisfactory method is an upward-flow type with provision for backwashing. It may be preferable to filter the flow to each hatchery trough separately. Purpose-made filters are available but a simple

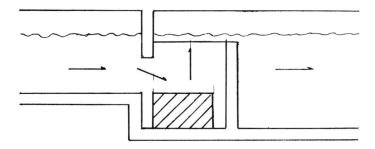

Fig. 10.1 Diagram of horizontal upward-flow screen in the intake channel to earth ponds

Fig. 10.2 Prefabricated fish-screen for 'monk' pond outlet

arrangement made from a catering-size coffee can or similar container, filled with limestone chippings, works well and can be replaced for cleaning.

Hatchery equipment   Where large numbers of eggs have to be counted, counting is done with an electronic counter. Most of these machines will

Fig. 10.3 Exterior fish screen for tank outlet

also recognize, count and discard dead eggs. Eggs can also be counted with reasonable accuracy by simple displacement, but 'green' eggs must have had time to swell to their full size.

Automatic egg pickers are available, but most hatchery managers prefer to take out the dead eggs with an ordinary suction bulb when they can be counted and recorded. Large numbers of eggs incubated in several layers or in flasks have to be treated with a fungicide.

The fry in a hatchery tank can be counted by weight. Several samples, each of at least 20 fish, are netted, using a muslin fry net, and transferred to a small perforated plastic container attached to an accurate balance and weighed. The fry in each sample are counted and an average weight for the fish in the tank can be worked out. Netloads of fish are then caught up and poured into water in a large container, such as a plastic dustbin standing on an accurate weighing machine. The increase in weight is read from the scale and the number of fry is worked out from their average weight.

Heating water   A source of heat to warm surface water can usefully accelerate the development of fry and fingerlings. It may also be economically worthwhile to use warmed water to speed up the growth of fish intended for on-growing in pumped salt water onshore or cages in the sea. The cost becomes higher as the fish grow and need a progressively increased flow. Electricity through a 9 kw immersion heater is the simplest way to raise the temperature of the water for incubating eggs and rearing fry in the first six to eight weeks after swim-up.

Figure 10.4 shows an unusual and highly practical way of warming up the borehole water used to supply the fry tanks on a Danish trout farm. Other methods are needed if the use of warmed water for fingerlings or pre-smolt steelhead parr is to be cost effective. A large trout farm can make profitable use of a heat pump (Fig. 10.5) to extract latent heat from the soil or from deep fresh water or the sea. The heat is then

Fig. 10.4 Warming the borehole water supply to Danish fry tanks inside the building

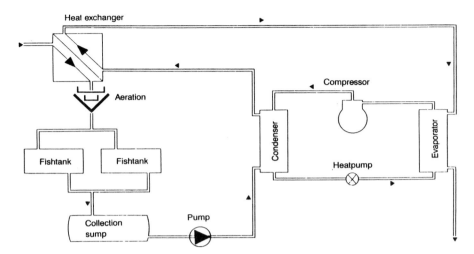

Fig. 10.5 Diagram of a heat pump water-warming system for fish rearing tanks designed by Sea Farm Trading of Bergen, Norway

transferred through a heat exchanger to the water circulating in the fish tanks.

Cooling water from electricity generating stations or factories has been used through heat exchangers to warm water for fish culture. The supply of waste heat is entirely dependent on the commercial operation of the industrial source. If the power station or factory shuts down the loss of heat can have disastrous consequences for a subordinate trout farm.

Fish handling Fish pumps and mechanical fish lifts can be used to shift the fish for transfer or grading (Fig. 10.6) Long-distance live-fish transport should be undertaken by specialists. The purpose of all mechanical gear is to avoid physical contact with the fish and to keep them inside a cushion of water. Any fish-handling equipment which fails to do this is worse than useless as it is likely to cause injuries.

Feeding Automatic feeding is normally required for the fish held in raceways, tanks or cages. It is essential for fry and fingerlings,

Fig. 10.6 Live-fish pump and mechanical grader by Milanese, Italy

which have to be fed at frequent intervals with a graduated amount of feed (Fig. 10.7). Hand feeding may prove more economical for fish in earth ponds and keeps the fish master aware of the condition of the fish.

Security The most common cause of loss in land-based trout farms is failure or contamination of the water supply. Sensors are available to monitor the chemistry, quantity and temperature of the water. The warning signals can be transmitted by radio or public telephone. The best protection against theft is light and the whole farm should be illuminated during the hours of darkness. Various devices can be had, including video recorders and radar, to intercept intruders, but they are expensive and far from foolproof. The best defence is a high, wire-mesh, continuous perimeter fence, topped with three strands of barbed wire, strung to lean outwards.

Fig. 10.7 Automatic feeders in fry tanks

Security for cages in fresh water is a problem that is best solved by specialists. Sea cages can be kept under continuous surveillance by radar but nothing can beat human supervision. The cage farmer's worst enemy is the best safeguard, bad weather.

Closed recycling This system involves the recycling of all the water in the whole unit. Not more than 2–3% of the total volume in circulation is replaced during 24 hours. The water is filtered and purified, chemically and biologically, oxygenated, and maintained at a stable, predetermined temperature. The whole unit can be under cover.

Full-scale units can be obtained from manufacturers as complete working installations. It is vital to assess their potential profitability. Loan charges on the total invested capital, together with depreciation and a guaranteed estimate of running costs, including labour, must be compared to potential production and realistic value of sales.

There could be some long-term economic advantages in a recycling unit producing 200–300 t of trout close to a large urban area. The small quantity of make-up water needed, together with the correspondingly small volume of effluent, might lead to a choice of site where development could go ahead without expensive restrictions.

Profitable design Rainbow trout in a current of water are obliged to swim in order to maintain station. In static water they swim at will in pursuit of prey, to avoid predators and to keep contact with other fish of their own species. Swimming speed is proportional to length, but it is very difficult to determine either the idling or maximum speed of any size of trout. Both adults and juveniles have been tested in float channels, where the opposing current can be progressively increased, but it is impossible to tell whether they deliberately turn and swim downstream, or fall back because they are exhausted. Farmed fish kept in raceways, or tanks with a concentric flow, must swim to maintain station. In doing so they are using energy sources which might otherwise have been stored in growing body tissues. In terms of conversion and growth rate rainbow trout do well in static water where there is no directional flow, but they need more living space and consequently have to be kept at a much reduced density. Where the available area is small they do better in tanks or raceways with the minimum speed of flowing water necessary to supply sufficient oxygen. Given adequate level, non-porous ground, ponds of the Danish type with a fairly rapid interchange of water were a good economic proposition. At present the most profitable system of on-growing rainbow trout is to use floating cages where an interchange of water takes place without excavation or channelling, with a pumped or gravity supply, and where the fish can get sufficient oxygen without having to swim actively against a current.

11 Trout food and feeding

Trout are carnivorous fish and exist by catching and eating other living organisms. Their digestive system is designed to handle animal protein and they can only digest and make use of a strictly limited variety of vegetable products. Most of the fish food mixtures originally given to domestic rainbow trout, reared for the table market, were arrived at through common sense allied to trial and error. Scientific research into the dietary requirements of rainbow trout has been mainly undertaken in the USA, and we have to thank intensive American research work for our real knowledge of the basic nutrient requirements of rainbow trout.

Basic diet
Protein

For practical purposes it is true to say that the best rainbow trout foods are those which contain the maximum amount of animal protein. A low grade food might contain 28−35% and a high grade food 45−50%. The total protein content of most food mixtures is made up by the addition of vegetable protein. This can only be utilized by the fish in comparatively small quantities and, if fed in large amounts, can be actively detrimental. Table 11.1 shows the suggested certain minimum amino-acid requirements for salmonid fish (not necessarily rainbow trout).

Carbo-hydrates

Trout can utilize small quantities of digestible carbohydrates (glucose, lactose etc), but not more than 9% of digestible carbohydrate should be given to trout, and the daily intake should not exceed 4.5 g of digestible carbohydrate per kilogram weight of fish. If too much digestible carbohydrate is given for any length of time, heavy losses will occur. The bodies of the dead fish will be seen to be swollen and, when opened up, the livers will be grossly increased in size and light in colour; this is due to overstorage of glycogen. Carbohydrate, in the various kinds of grain meal often mixed in trout foods, can be

Table 11.1 Minimum amino-acid requirements for salmonids

Amino acid	% of diet
Arginine	2.5
Histidine	0.7
Lysine	2.1
Methionine	0.5
Cystine	1.0
Tryptophan	0.2
Threonine	0.8
Valine	1.5
Leucine	1.0
Isoleucine	1.5

given in fairly large quantities as it is practically indigestible by trout and therefore does little harm.

Fats   A small amount of digestible fat is necessary in the trout diet. The digestibility of fats depends upon their having a low melting point as they must be liquid in the stomach of the fish in order to be utilized. The essential fatty acids are linoleic, linolenic and arachidonic, all of which are higher unsaturated fatty acids. A practical food mixture will contain about 5–8% of fat. Too much fat in the diet results in fish losses due to fatty degeneration of the liver and kidneys.

Minerals   Small quantities of minerals are as essential to trout as they are to higher animals. The quantities are not known, but it can be assumed that most of the minerals essential to life can be taken up by the fish directly from the water. A fish's body is composed of 70–75% water and water is an essential nutrient. If an analysis of the water on a freshwater trout farm indicates a natural poverty of minerals, an agricultural mineral mixture can be added to form up to 2% of the food intake. Sea-salt containing iodine has been found to be a beneficial addition to trout food mixtures and can be given in a proportion of up to 4% of the food intake. A trace of iodine (0.0006–0.0011 mg per kilogram of live body weight of fish) is regarded as essential in the trout diet.

Vitamins The daily vitamin requirement has been suggested by research workers in Europe and the USA. Table 11.2 lists the essential daily minimum quantities per kilogram of the live body weight of the fish.

Table 11.2 Daily vitamin requirements

Vitamin	Minimum quantity (mg)
Thiamine (B_1)	0.150−0.20
Riboflavin (B_2)	0.50−1.0
Pyridoxine (B_6)	0.25−0.50
Biotin (H)	0.04−0.08
Nicotinic acid	4.0−7.0
Pantothenic acid	1.0−2.0
Folic acid	0.10−0.15
Inositol	18−20
Choline	50−60
Cyanocobalamin (B_{12})	0.0002−0.0003 (a trace only)

It is now known that trout require vitamin C. They may also require the fat-soluble vitamins (A, D, E and K) and these are usually included in most commercial dry feeds (Table 11.3)

Table 11.3 Additional vitamin requirements

Vitamin	Minimum quantity
Vitamin A	8000−10000 i.u.
Vitamin D	1000 i.u.
Vitamin E	125 i.u.
Vitamin K_3	15−20 mg
Vitamin C	450−500 mg

Calorie requirements of trout Relatively little is known, as yet, of the essential calorie requirements of rainbow trout. Research is very difficult because the fish are poikilothermic, and it is consequently hard to relate changes in body temperature to the utilization of food. Trout have little ability to utilize carbohydrates and fats as sources of energy and most of the energy potential of trout food must therefore come from the protein content. Rainbow trout are only capable of utilizing about 15% of vegetable protein so that, in practice, nearly all the metabolic energy output is derived from the animal protein in a diet. Some manufacturers of dry food quote a figure for the metabolic energy potential of their food formulas (ME—kcal/kg). This must be a nebulous figure as a measure of the value of the particular food is concerned, as the calorie requirements of rainbow trout are not fully understood.

It is frequently claimed by manufacturers of dry foods that a particular food formula, with an allegedly high energy content as a fish food, produces a low (better) conversion rate. This is self-evident as the best trout foods, i.e. those with the lowest and most efficient conversion rates, are the ones which contain the highest percentage of good-quality, digestible animal protein, and these are the foods which automatically have the highest energy content.

'Dry' feeds for rainbow trout

American research was responsible for working out the basic nutrients needed for trout, and the development of most of the dry foods for salmonid fish, particularly rainbow trout, is founded on the investigations originally made at the Cortland Laboratories in the USA.

Dry fish food is a formulated compound which can be fed to rainbow trout of all ages and sizes, either directly as a whole diet, or when added to fresh food. It comes in the form of fine crumbs, which are fed to fry, and different grades of pellets, suitable for growing trout of larger size. The commercial dry foods are 'secret' in the sense that they are branded and made to a formula which may have been devised by a particular manufacturer. It is, however, possible although usually uneconomic, for any competent miller to produce a dry fish food, if the mill has pelleting equipment.

The main effective constituent of all commercially prepared fish foods is dried animal protein. The better the food, the more good-quality animal protein it will contain. The best source of suitable animal protein is high-grade fish meal. This should be vacuum dried and derived from whitefish. Poor quality, heat-dried fish meal, particularly if derived from fish of the herring family, can lead to dietary deficiencies and to the death of large numbers of fish on a fish farm, if it is fed for any length of time as a total diet.

Commercial feeds

Commercial dry rainbow trout foods are usually made in the following grades and qualities:

1 *Starting and fry feeding*. Three grades of crumbs.
Usually made up with more protein and more animal protein than pellets for larger fish. A reasonably good fry food should have a protein content of about 50%, of which at least 75% should be animal protein. Used for trout up to a size of 5 cm (670 fish to the kilogram).

2 *Growing pellets*. Two or three grades. The protein content should be about 45%, of which at least 70% is animal protein. Used for trout from 5 to 15 cm (670−25 fish to the kilogram).

3 *Finishing pellets*. Two grades. Protein content should be 40−45% of which about 60% is animal protein. For fish up to slaughter size.

Special, high-protein pellets are manufactured for brood fish.

The other constituents of commercially prepared foods are digestible fats of low melting point (which usually make up about 7−8% of fry foods and 5−6% of foods for larger fish) and added vitamins and minerals.

Unless the manufacturer can guarantee that no herring meal is used in food, vitamin B_1 (thiamine) should be added during manufacture, and the storage life of the vitamin should be stated on the bags.

Most manufacturers provide special pellets which will produce a red colour in the flesh, when fed to the fish for the last four to eight weeks before they are slaughtered.

Water content

Dry foods are not really dry in the true sense of the word. It is particularly important that the initial moisture content should not exceed about 10% and that the food should be kept in a cool, dry store. Pellets and meals can be attacked by fungi which produce by-products that are directly toxic to the fish.

Quantities of food and conversion rates

The amount of food required by rainbow trout is directly dependent on water temperature. Most manufacturers of dry foods provide a table showing the daily intake of food required by the different sizes of rainbow trout, at temperatures ranging from 5 to 20°C.

Dry food conversion

It is usually claimed by the manufacturers that their dry foods can achieve a conversion rate of 1 to 1.1, under optimum conditions. It is certainly possible to achieve a conversion rate of 1 to 1.4 with a good-quality dry food, using carefully-regulated automatic feeders. It is unlikely, however, that a better overall conversion than something in the order of 1 to 1.5 is obtainable when the fish are in Danish-type earth ponds or large tanks, owing to waste. Trout will seldom take pellets off the bottom of an earth pond, particularly if it is muddy, although they will do so in a pond with a concrete bottom.

The conversion rate of a commercial fish feed is termed its coefficient by the manufacturers and is stated in decimals ranging from 1.1 for the best grades, containing a maximum proportion of animal protein, down to 1.7 or greater for the poorer grades.

Moist pellets

Ordinary 'dry' pellet fish feeds have a moisture content of about 12%. Some fish farmers believe that their fish are healthier in fresh water as well as in the sea when they are fed with moist pellets with a moisture content of 20−50% of their total weight.

Moist pellets are normally made up on the farm. They consist of a dry component, usually a vacuum-dried fish meal, and other protein meals with added vitamins, low-melting point fat usually in the form of fish oil, an anti-oxidant and a binding agent. The binding agent most commonly used is methyl cellulose. The binding agent absorbs water quickly and forms a stable, soluble gel. About 1% is added to the food mixture and is usually supplied as part of vitamin meal.

Hygroscopic pellets Making up moist pellets from dry components involves the problem of storing the separate ingredients. It is uneconomic to bring small quantities to isolated sea farms and large consignments may deteriorate before they can be used. On-site production is consequently only a good proposition for large units.

A recent development has been the commercial manufacture of a pellet food which can be stored dry but which rapidly absorbs water if wetted before it is fed to the fish. The pellets swell but do not disintegrate as might be expected when they are sprayed with water. The binding agent used is mixed dry in the pellet. When wetted it rapidly absorbs water to form a stable gel which coats the other components of each separate pellet. The wetted pellets must be fed to the fish straight away and cannot be stored. This means that a fresh batch of food has to be made up for each feed and fed to the fish by hand.

Fish silage An animal-protein silage can be made from minced, raw industrial fish or by using the fish waste from a processing plant. Fish silage can form a valuable food for on-growing large rainbow trout in the sea. It is easy to make and can be stored for long periods without refrigeration but is only economically worthwhile if regular supplies of the raw materials can be obtained close to the sea farm.

Preparation Fish silage is made by adding 3–4% of an acid to the minced fish, which must be fresh either from the sea or from cold storage. The acid can be inorganic or organic. Formic acid is commonly used. The pH of the mixture is brought down to below 4. This inhibits bacterial decay but the 'digesting' process by enzymes in the minced fish guts goes on and reduces the mixture to a liquid slurry. An anti-oxidant is added to prevent the fats becoming rancid and the liquid can be stored in tanks for up to six months.

Use A 50/50 mixture of silage and a commercial feeder meal containing vitamins and a binding agent is passed from silos into a simple perforated-plate extruder. A moist pellet made from high-fat industrial fish will contain 17–18% low-melting point fats. The oil content can be increased to 24% by adding fish oil to the mixture before extrusion.

The fish weight-gain cost from moist pellets made from fish silage can be 9–10% lower than for the equivalent 'dry' pellet food. The moist pellets cannot be stored for more than a few days and should be used within 24 h in hot weather.

Times of feeding

The best results are obtained by feeding rainbow trout little and often. Automatic feeding is the answer, as the timing of feeds, and the quantity of food given, is open to a wide range of adjustment. When hand feeding has to be employed, with either dry or fresh, wet food, fry should be fed at least six times a day during the first four to five weeks of feeding. The feeds can then be reduced to five a day for the rest of the first summer's feeding period. Large fish, in the second summer, should be fed two to three times a day, dependent on the water temperature. Trout should not be fed in earth ponds after 9 a.m. in summer, if the oxygen content of the water is low.

Autofeeding times

Automatic feeders should be regulated to deliver the daily food requirement, according to the weight of fish in the pond or tank, over the maximum number of hours of daylight. The system can be adjusted to exclude periods of the day when the oxygen level is low, or the water temperature is above 21°C. Automatic self-feeders have the advantage that the fish food is not available haphazardly, but only when the fish are hungry.

Growth in fresh water

The growing times of rainbow trout, cultured in fresh water for the table market, are usually divided into feeding periods. These are the periods during which the trout are feeding actively and making effective growth. In the northern temperate zone this period is approximately from early summer to autumn, but it can be extended for a few weeks in the south and progressively shortened towards the north.

Sexual maturity The majority of the rainbow trout remaining on a freshwater farm will become sexually mature at the end of the third year and, unless a late maturing stock has been established, it is seldom worthwhile carrying fish which are intended for the table market over a third winter.

Flesh colour of rainbow trout It has been known for a very long time that the red colour of the flesh of some wild salmonid fish is due to the presence of a fat-solvent pigment of the carotenoid group. The fish collect the pigment by eating other animals which have assimilated it from their food.

It is possible to produce a red colour in the flesh of rainbow trout by feeding crustacea, either dried or preferably fresh, to the fish, for a period of about four to five weeks before they are slaughtered. Fresh prawns or prawn and shrimp meal is expensive, and is only worthwhile incorporating in the fish food when a correspondingly high price can be obtained for the fish, or when the crustacean food can be obtained very cheaply.

Synthetic additives Carotenoids have been added to the feeds of animals and poultry to colour the food products derived from them. The natural carotenoid pigment present in most salmonid fish species is a xanthin. A similar red-coloured carotenoid pigment known as canthaxanthin has been synthesized and is now produced commercially. It has been approved for food colouring use in Europe and Canada. Canthaxanthin fed to rainbow trout at a rate of 190 mg/kg of food produces a satisfactory red colour over a period of approximately 10 weeks. A deeper red colour, approximating to that of salmon, was produced by feeding the pigment at a rate of 450 mg/kg of food for a period of seven weeks.

It has been suggested that canthaxanthin could be a carcinogen although there is very little supportive evidence. An alternative natural carotenoid is available, known as astaxanthin.

Dietary deficiencies   The commonest cause of dietary trouble in rainbow trout is using a cheap feed, low in animal protein (lacking essential amino acids). Another deficiency which often causes large losses is a lack of thiamine (vitamin B_1), in either wet or dry feeds which contain too much herring.

Symptoms   The following symptoms have been attributed to a lack of particular vitamins in the diet of rainbow trout. How far these can all be substantiated in practical trout farming, as well as in the laboratory, is not yet established.

Thiamine (B_1) Loss of appetite; instability and impaired equilibrium; convulsions before death. The treatment is to make up a $1-2\%$ aqueous solution of thiamine and to spray this on the dry feed (or mixed in a wet food diet) at a rate of $1l/50$ kg of food.

Riboflavin (B_2) Loss of appetite; fish seek shade or darkness and swim deep in ponds; vision seems impaired; eye lens may be clouded and eyes bloodshot; the fish become dark in colour.

Pyridoxine Loss of appetite; overactivity and nervous reaction; rapid breathing and gasping mouth movements; quivering of the gill covers; fluid collects in the body cavity; the fish are anaemic and the skin on the back may darken; bodies stiffen almost immediately after death.

Biotin (H) Loss of appetite; muscular atrophy and convulsive movements; darkening of the skin; sores in the intestine.

Nicotinic acid Loss of appetite; movement becomes spasmodic and jerky; fluid collects in the stomach and intestines.

Pantothenic acid Loss of appetite; fish are generally unhealthy looking; gill filaments may be stuck together and covered with mucus; sores may appear on the body.

Folic acid Slow growth; the fish are sluggish; anaemia; colour darkens; the fins, particularly the tail fin, have a broken appearance.

Inositol Poor growth and conversion; distended stomachs.

Choline Poor growth and conversion; fatty degeneration of the liver; bleeding in the kidney and intestine.

Vitamin E Poor growth and conversion; darkening of the skin.

Controlling growth

The most important know-how which distinguishes the real fish master is not the ability to grow trout quickly and economically, but to control their rate of growth and the time at which they reach market size. The best trout farms have trout ready for the market each week throughout the year. The knowledge necessary to bring trout on at the right time can only be acquired by practical experience, but every good trout farmer has to find out how to do this in the particular circumstances which apply on his own farm.

Feed cost

The salmonids, in common with nearly all the other coldwater fish, are entirely carnivorous. Any carbohydrates they may inadvertently eat are in the guts of their prey. Their diet consists entirely of protein, and the fats associated with it in the tissues of other poikilothermic animals, particularly other fish. The fats in these animals are low-melting point and digestible. Farm feed mixtures cannot be safely compounded from mammalian tissues where the protein is associated with indigestible, solid fats. Vegetable proteins and oils have been used but only as additives.

Nothing has, so far, been found to be a satisfactory substitute for the main constituent of fish-farm feed which is fish meal, principally derived from industrial fish. Useful species such as capelin (*Mallotus villosus*), large and small sandeels (*Hyperoplus lanceolatus* and *Ammodytes lancea*) and sprats (*Sprattus*

sprattus) are in short supply and are being exploited almost to extinction by uncontrolled over-fishing. The price of fish meal is rising and the quality is falling. The long-term future for farming any salmonid fish may depend on finding some satisfactory diet from lower down the food chain.

Pollution hazard

The development of high energy feeds by manufacturers has, to some extent, reduced the pollution from trout farms due to the products of metabolism in the fish. The increased nutritional value of the food results in a corresponding reduction in FCR, or feed conversion ratio, which is usually taken as the weight of feed required to produce 1 kg of fish.

The so-called pollution hazard is the difference between the nitrogen and phosphorus in the nutrient in the feed and that retained by metabolism in the growing tissues of the fish. The nitrogen content of rainbow trout tissue can be taken as 3% and the phosphorus 0.5% of their total weight. The nitrogen content of a particular feed can be derived by dividing the weight of crude protein in that feed by 6.25.

For example, if the total phosphorus content of a particular feed is known to be 0.9%, then production of 1000 kg of rainbow trout, given 1000 kg of feed with an FCR of 1.0, containing 420 kg of crude protein and 9 kg of phosphorus, will create a pollution hazard of 37.2 kg of protein and 4 kg of phosphorus.

12 Common diseases – recognition, treatment and prevention

Nearly all the diseases which can occur in epidemic form among rainbow trout on fish farms are indirectly attributable to the domestication of the fish, and the density at which they are kept when reared for the table market. Wild trout stocks, in rivers and lakes, very seldom suffer the massive losses which can occur on fish farms.

Diseases in fish may result from any of the following conditions:

Bacterial or viral infection;

Infestation by internal or external parasites;

Environmental conditions such as lack of oxygen, entrained gases in the water or physical damage following skin abrasion or gill clogging;

Toxic algal blooms;

Deficiencies or toxins in the diet.

Some disease pathogens are present only in fresh water, some in the sea and others in both fresh and salt water. The sea-going salmonids are doubly at risk. Diseases can be transferred from fresh to salt water within the young fish, or the pathogenic effects of a disease which infected the fish while in fresh water may become apparent when they are stressed on removal to the sea.

Fish pathogens can be separated into two main groups. Those which are termed obligate are normally absent from water in which there are no diseased fish or carriers of disease. Many of the common bacterial and viral diseases belong to this group. The second group is termed facultative. These are pathogens which are naturally present in the water and may infect fish and cause symptoms of disease when they are stressed or there are physical changes in their environment, such as abnormal fluctuations in temperature or salinity.

BACTERIAL DISEASES

Furunculosis The most common of all bacterial diseases on trout farms. It
(causative can occur in fresh or salt water, and infection is most likely at
organism higher temperatures, 15–18°C or above. The disease is spread
Aeromonas in the water, or by direct contact. Infected eggs or fish car-
salmonicida) riers are also sources of the spread of infection. Rainbow
trout can carry the disease without exhibiting pathogenic
symptoms. The disease is endemic on many trout farms, but
overstocking and bad hygiene contribute to epizootic outbreaks
which can cause heavy losses.

Symptoms The disease develops after an incubation period of three to
four days and fish can die in large numbers without exhibiting
any characteristic symptoms. Sub-acute attacks produce in-
flammation of the intestine and a reddening of the fins. The
most typical symptom of the disease in acute form, which has
given rise to the name, consists of swellings or furuncles that
can occur anywhere on the fish's body. The boils contain a
reddish pus formed of dead tissue. The fish's pectoral fins are
usually infected and the fin tissue dies. Eventually the whole
fin can disappear. Stumps of pectoral fins are an indication
that fish have recovered from furunculosis.

Treatment If the fish continue to eat, the disease can be treated with an
antibiotic mixed with the food. The chemical compound most
commonly used has been oxolinic acid made up as a commercial preparation of 1 g oxolinic acid per 5 g of powder
base. This given for 10 days at a daily rate of 50 mg/kg of
body weight of fish. The bacterium has developed an increasing degree of resistance to oxolinic acid and other methods of
treating this disease are being used. These include a return to
sulpha drugs as well as more general use of oxytetracycline.
Furazolidone has tended to replace sulphamerazine in the
treatment of furunculosis. Some 10–20 mg/kg of live body
weight of fish is given prophylactically and the therapeutic
dose is 50–100 mg/kg of fish.

It is particularly important to remember that the residues of chemicals in the tissues of fish treated for bacterial infection may be harmful to human beings. The fish normally eliminate harmful concentrations over a period of 10–12 days at temperatures of 10–15°C. The period taken for elimination in colder water – between 4 and 10°C – is much longer and may exceed 30 days.

Aeromonas liquefaciens This bacterium belongs to the same general group as that causing furunculosis. It is becoming increasingly common on rainbow trout farms and can occur in both fresh and salt water.

Symptoms These are not unlike those produced by furunculosis. Surface lesions on the fish's body are smaller and tend to break down into open sores before swellings can form. Fin tissue becomes reddish in colour and breaks down.

Treatment The disease responds to the same treatment, given either as a prevention or cure, as that for furunculosis.

Vibriosis This disease is a haemorrhagic septicaemia caused by the bacterium *Vibrio anguillarum* (and possibly other *Vibrio* spp.) which is present world-wide, generally in marine or estuarial environments. It can infect salmonids in fresh and salt water but outbreaks appear to be more frequent and damaging in some areas than in others and in waters which reach comparatively high temperatures.

Symptoms The fish cease to feed and become lethargic. Haemorrhagic areas appear in the skin and there is a reddening at the roots of the fins, the vent and sometimes in the mouth. Bleeding occurs in the gills and intestine. Deep red sores may appear on the body. Outbreaks of the disease in acute form, such as can occur among young pink or chum salmon reared in salt water in shore-based tanks, may produce no external symptoms other than large-scale mortalities.

Treatment   Similar to furunculosis. A vaccine given by immersion has provided a measure of protection, and the fish can be treated prophylactically with trimethoprim.

Bacterial kidney disease (BKD)   This disease is caused by a *Corynebacterium* spp. and affects salmonids in Europe and North America in both fresh and salt water. It can be carried over from fresh to salt water. Previous infection during freshwater life can cause large losses as fish with damaged kidneys fail to stand up to the osmotic stress of transfer to the sea. In salt water the disease generally appears during the first winter.

Symptoms   Chronic infections of young fish in fresh water may go unnoticed until casualties occur after the fish are transferred to the sea. Fish which die will be found on dissection to have whitish lesions in the kidney and bleeding from the kidney and liver. Infected fish in sea cages may cease to feed and swim near the surface. They can appear dark in colour when viewed from above and show swellings on the sides. The eyes may bulge. There can often be no external symptoms and large fish in salt water may continue feeding actively, until they suddenly die for no apparent reason.

Treatment   Similar to furunculosis.

Bacterial gill disease   The causative organism of this disease is a *myxobacterium*. The disease is common among all salmonid fish and is most likely to attack rainbow trout in the fry stage.

Symptoms   In the initial stages of the disease the fish appear lethargic and have a poor appetite. An examination of the gills and gill filaments will show that they are swollen and a deeper red colour than normal. At a later stage the gill filaments become clumped together and pale in colour. The most characteristic symptom of the disease in the later stages is a heavy secretion which clogs the gills and can be seen exuding from the gill covers.

Treatment   Avoid overcrowding. Filter the water supply to remove irritant debris and shake out the dust from starter feed using a very fine mesh domestic sieve. Short-term baths in a solution of a bacteriocide such as chloramine T or benzalkonium chloride (BKC) may be used. The recommended concentrations vary in relation to pH and reference should be made to the appropriate manufacturer's instructions.

PROTOZOAN DISEASES

Whirling disease (myxosomiasis)   This disease is caused by a spore-forming parasite called *Myxosoma cerebralis*. The latter part of the name is derived from the behaviour of the parasite which, during the free-swimming stage, travels through the fish to become encysted in the area of the brain. The life cycle is not fully understood but spores, which are shed into the water or come from dead fish, are present in infected ponds and appear to be ingested by the live fish. The spore capsule is dissolved in digestive juices, releasing the free-swimming protozoan into the gut. The protozoan then finds its way into the fish's bloodstream and migrates to the area of the brain where it forms a cyst. Trout are attacked in the fry stage and the protozoa are no longer able to form cysts after the early, cartilaginous structures have become ossified.

Symptoms   The spores can be dormant for years in a dry pond. Infection takes place as soon as trout fry are released into the infected water, and the fish begin to exhibit symptoms after a lapse of about 40–60 days. The first signs of the disease are that some of the fish will show a darkening of the skin covering the tail fin and shortly afterwards will be seen spinning round, chasing their own tails. At this stage heavy mortalities will occur. The cause is through damage to the balance mechanism in the inner ear and to the sympathetic nervous system which regulates pigmentation of the skin.

Fish which have survived the disease are frequently deformed or dark in colour. Deformities can take the form of irregular foreshortening of the gill covers and distortion of the vertebrae, particularly at the root of the tail fin.

Treatment   There is no treatment for the disease, but it can be eliminated in the trout on a trout farm without much difficulty. Fry must be kept in a spore-free water until ossification is complete. This occurs after about 10−12 weeks feeding, when the fish have reached a size of 5−8 cm. The fry must not be put into earth ponds before ossification, but should be kept in tanks. The water supply to fry tanks should be drawn from springs, or from a source that contains no fish.

Infected ponds should be treated with calcium cyanamide $(Ca (CN)_2)$, at a rate of 0.5−0.75 kg m^2 when the ponds are dried out. Care must be exercised to avoid pollution of waters containing live fish when ponds are refilled.

Costiasis   The disease is due to a protozoan parasite (*Costia necatrix*) (Fig. 12.1) which attaches itself to the skin or gills of the fish. The parasites can be seen quite easily under a low-power

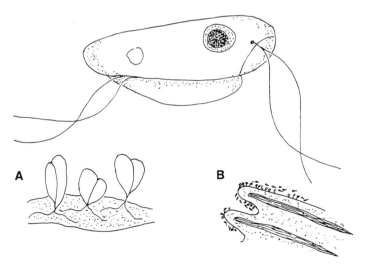

Fig. 12.1 Costia necatrix (can occur in other forms) A, Parasites attached to epithelium; B, Appearance of skin in badly infected fish

microscope. Infected fish have what looks like a blueish-grey layer of slime on the skin; when this is scraped off and examined, it can be seen to consist of a mass of parasites. The protozoan multiplies by simple division and can probably exist in dry conditions by forming a capsule. The parasites quickly detach themselves after a fish has died, and very few will be found on the body after a lapse of about an hour.

Treatment A formalin bath. See page 130.

Hexamitaisis Octomitis (*octomitis salmonis*) This disease is caused by the presence of a pear-shaped microscopic protozoan (*Hexamita truttae*) (Fig. 12.2) in the gut of trout. The organism multiplies by simple division and will form a capsule under adverse conditions. The disease can cause considerable losses among fry and small fish.

Symptoms The disease causes steady rather than sudden losses and the affected fish first become lethargic, then appear suddenly to lose their ability to orientate and make random movements, eventually sinking to the bottom. Some fish will be seen to 'flash' (turn about the axis of their bodies).

Treatment The treatment is simple and consists of feeding Calomel, 0.2% with the fish food, for a period of four days.

White spot disease The causative organism of this disease is a single-celled, ciliated protozoan (*Ichthyophthirius multifilis*) which can be seen with the naked eye; fully grown specimens are about 1 mm in diameter. The parasites are roughly circular or oval in shape and covered in cilia (hairs). They can be seen swimming over the slime layer in heavily infested fish.

Symptoms Clearly defined white spots, which can join together to form irregular white patches on the body. The fish can often be seen trying to rid themselves of the parasites by rubbing on the bottoms or the sides of ponds and tanks. Heavily infested fish become lethargic.

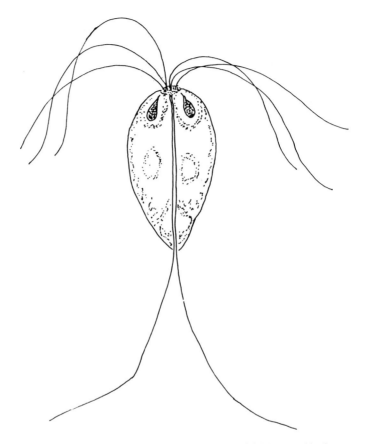

Fig. 12.2 Octomitus salmonis (Hexamita truttae) (highly magnified)
(After Davis)

Treatment   The parasites are not only present on the surface but can dig
themselves into the skin, where they are protected by living
tissue. The surface parasites can be killed off by a formalin
bath. See page 130. More deep-seated parasites may respond
to bathing with solutions of copper sulphate, malachite green
or common salt.

The best way to prevent infestation is to keep the fish in
fast-flowing water. The parasites will not tolerate drying out

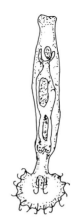

Fig. 12.3 Gyrodactylus spp.
Common ectoparasites of
trout (magnified about 70 times)

and a pond can be disinfected merely by being laid dry for a period.

DISEASES CAUSED BY FLUKES (TREMATODE WORMS)

Gyrodactylus   One of the most common parasites on trout is a small, transparent fluke belonging to this family. It is equipped with two hooks at the front end with which it attaches itself to the skin of a fish. Massive infestations can cause heavy losses. *Gyrodactylus* produces live young which attach themselves to the body of the host animal as soon as they are born.

Symptoms   The parasite can be found anywhere on the fish's body, but is usually concentrated on the back and tail fins. The area of skin, where parasites are attached, becomes covered by a blueish-grey layer of slime. Areas of skin, on both fins and body, can be eaten away, and the lesions become infected with fungus.

Treatment   *Gyrodactylus* can easily be seen with a hand lens, in a sample of slime taken from an infected area of a fish. Treatment consists of a bath lasting 1 h in a solution of formalin. See page 130.

Eyefluke
(*Diplostomum*
spataceum)

A stage in the life cycle of this parasite fluke causes partial or complete blindness in trout, and presents a serious problem to the rainbow trout farmer. The sexually mature animal, which measures 2.3−4 mm, lives in the gut of several species of gull. The mature fluke lays eggs, which are shed in the gull's excrement, and may fall into water. The fluke eggs in the water develop into a larval stage during which they become attached to species of freshwater snail. The larvae penetrate the snail's body and migrate to the liver, where they multiply for three generations. The final generation develops into a microscopic, free-swimming form, with a divided tail (furcocercariae) (Fig. 12.4).

The free-swimming larvae that come in contact with a fish penetrate the skin or enter the gill filaments. Once inside the fish's body the larvae shed the twin tails, becoming oval in shape (metacercariae). In this form they migrate to the eye lenses of the fish, where they cause progressively increasing blindness. The degree of infestation is measured by counting the number of parasites present in each eye lens, and can range from four or five up to more than 100. The larvae remain active for at least eight months.

Symptoms

Infested fish have cloudy eye lenses, and the fluke larvae can be easily seen in a preparation made from eye-lens tissue, under a low-power microscope.

Infected fish do not die, but find it progressively more difficult to obtain their proper share of food, and their general condition deteriorates until to keep them alive becomes a liability.

A trout farm producing solely for the table market can tolerate a low level of infestation with eye fluke, as the fish manage to feed reasonably well and maintain growth up to slaughter at 'portion' size; although they may not grow as well as fish with undamaged eyesight. The damage is irreversible and partially blinded fish are very undesirable as brood stock, or for further rearing in salt water. They are also of no use as stock for put-and-take fishing.

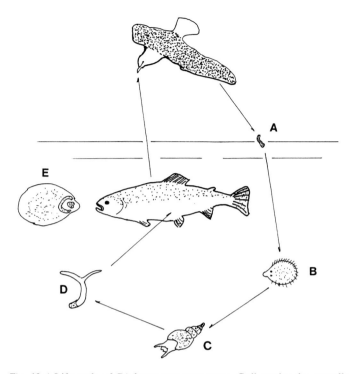

Fig. 12.4 Life cycle of *Diplostomum spataceum*. Gull carries the sexually reproductive stage of the parasite. Miracidia (A and B) are shed into water and invade snail host (C). Development proceeds asexually and free swimming. Furcocercariae (D) are released and enter a trout. They change to *metacercariae* (E) and concentrate in the fish's eyes. Fish is eaten by a gull. After Schäperchaus

Treatment Infested fish cannot be treated in any way and it is very difficult to prevent infestation taking place, if the water supply to a freshwater trout farm contains snails that are hosts to the parasite. The fish ponds and any water courses or channels leading into the fish farm should be cleaned periodically and kept as free from snails as possible. Ponds containing fish can be treated with copper sulphate solution. The concentration required to kill the snails is only slightly less than the level toxic to trout; consequently great care must be exercised to ensure proper mixing. This is very difficult in a large pond.

A method of killing eyefluke during its free-swimming stage in the water supply to trout farms has been developed in the USA. A simple electrode system is suspended in the intake channel and given a loading of 440 volts per inch from a mains supply, at 110 v or 240 v, through an isolating transformer. This electrode system is claimed to effect a kill of 80% of all cercariae carried downstream with the water supply. The parasite causing eyefluke and the various stages of its life history are described earlier.

VIRUS DISEASES

Infective pancreatic necrosis (IPN)
This disease and the causative virus is well known in the USA, and it appears to have spread from that country to Europe, probably with the importation of rainbow trout eggs. The IPN virus seems to have been introduced to Danish trout with foreign eyed eggs imported in 1960. Fry, and subsequently brood stock, derived from these eggs had no recognized mortality due to IPN, neither had the fry and brood stock in the second generation. The third generation of fry, which appeared in 1967 and 1968, were seriously affected by the disease. IPN has now spread to most European countries.

The highest mortality occurs among very young fish (fry) with a death rate of up to 85%. The fish which survive the disease retain the infection and become carriers, most likely for life. The general mode of transmission is by contact, probably through ingestion, and fry which survive an epidemic, remain infected and shed the virus with their faeces and, later on, with ovarian fluid and sperm. Disinfection of eggs does not prevent transmission of the disease, because the virus can be present inside the eggs.

Symptoms   Erratic swimming; corkscrew turning about the fish's own body axis (flashing); fish tend to go to the bottom before death.

Control   There is no treatment for the disease and the only method of control is by isolation and destruction of infected stock.

Infective haematopoitic necrosis (IHN)   This is a relatively new salmonid virus which was first isolated from young rainbow trout in the USA during 1967. The units where the disease was first reported experienced a loss of approximately 90% of fish over a period of three to five months.

Symptoms   Fish first become lethargic then swim erratically with lateral and longitudinal rotation (flashing), finally the fry float upside down, breathing rapidly and eventually die. Frenzied swimming is typical of the disease.

External symptoms   Earliest signs are long, opaque faecal casts, trailing from the vents and seen floating on the water surface, or collecting on the outlet screens; eyes bulge in their socket, areas of bleeding occur at the base of the pectoral fins, and on the body surface round the dorsal fin and vent.

Internal symptoms   Fluid in the body cavity; spots of bleeding and larger areas of bleeding in the wall of the body cavity and in the internal organs; liver, kidney and spleen pale in colour; among very young fish, there are signs of bleeding in the yolk sac, which is distended with fluid.

Control   In common with other salmonid virus diseases, there is no treatment and the only method of control is by isolation and destruction of infected stock.

Viral haemorrhagic septicaemia (VHS)   The disease, which is also known in Europe as the 'Egtved' disease or simply as the 'Virus', appeared on trout farms in western Germany in the early 1950s. It quickly spread to Denmark and then to most other European countries. The most common mode of infection is through live fish or infected equipment. There has been no instance of infection by eyed eggs, provided the water and packing material is not infected. It is a coldwater disease and outbreaks most frequently occur in winter. Summer outbreaks can also occur on farms using cold, spring water. Outbreaks are most likely when the fish have been subject to stress, such as that caused by grading or transportation.

Symptoms   Swollen abdomen; eyes bulging out of their sockets (Fig. 12.5); bleeding in or about the eyes; darkening of the skin; pale gills; heavily infected fish are lethargic and swim into shallow water, often with back fins above the surface.

Internal symptoms   Body cavity filled with a clear or yellowish fluid; liver swollen and yellowish grey in colour with broken blood vessels; kidneys swollen particularly towards the vent; flecks of blood in the lining of the body cavity; minor bleeding due to a breakdown of blood vessels is often general in the organs and muscles. Final diagnosis is only possible by viral investigation using tissue cultures.

Fig. 12.5 VHS infection on Danish trout farm. Fish with typical bulging eyes lying against outlet screen in back channel

Fish of all sizes are susceptible and the death rate in bad outbreaks can be up to 90–95%. In practice, no definite incubation period is discernible, although a period of one to two weeks has been observed in the laboratory.

Control The disease is spread downstream with the water from an infected farm and by the movement of live fish, or persons with infected clothing and equipment. The only method of controlling the disease is by isolation and destruction of infected stock.

DIAGNOSIS

Many different fish diseases produce symptoms of confusing similarity. For this reason it is particularly important that diagnosis is confirmed as soon as possible by appropriate tests carried out in the laboratory, in order that the correct treatment can be applied before it is too late.

OTHER AFFLICTIONS

Marine parasites The 'sea louse' (*Lepeophtheirus*) is a large copepod which lives on the skin of the host fish. The females are about 3–5 mm in length and are distinguished by a pair of egg sacs. The eggs are probably shed in the summer or autumn. Free-swiming nauplius and metanauplius stages are followed by the first chalimus stage when the copepod attaches itself to a host fish. Parasites in the chalimus stage can be seen on the fins of the fish and are about 1.5–3.0 mm in length. Several moults occur during the chalimus stages and are followed by numerous others before they attain their maximum size. The parasites feed on particles of skin and possibly blood drawn from the host fish.

Treatment Organophosphorous compounds are the most effective chemicals tested so far. The commercial preparations are powders soluble in water from which a stock solution can be made up. The manufacturer's instructions give directions for dilution and strength. Internal medication which renders the tissues of the host unattractive to the parasite has been tried out and may prove successful. Chemical treatment is difficult, particularly in large cages. Various methods have been tried. These include lifting the nets to reduce the enclosed volume of water and putting a polythene sheet completely under them or round the periphery to form a skirt. An added complication is that the fish are distressed by the medication and have to be prevented from injuring themselves or jumping out of the cage. Aeration may be needed while treatment is being given.

Danish research has indicated that *Lepeophtheirus* is unable to reproduce in salinities lower than 24–25‰. Rainbow trout can acclimatize much more easily to a marine environment in which the salinity is about 20–25‰. Water at the head of deep fjords or sea lochs, where a river runs into the sea, can have an average salinity of about 20‰. This is ideal for rainbow trout culture as the fish can be transferred to sea cages or enclosures without stress and remain free from sea lice.

Fish fungus   A common cause of loss in fresh water, particularly of eggs and alevin (yolk-sac fry) is infestation with *Saprolegnia* spp., the spores of which are present in the water.

Treatment   Malachite green (zinc-free) has been generally used to control *Saprolegnia*; either as a dip at 1:15 000 (67 ppm) with fish immersed for 30 seconds or as a bath (1:500 000 = 2 ppm) for fry and parr. The bath concentration can also be used as treatment for incubating eggs after they have reached the eyed stage.

Algae and dinoflagellates   Unlike wild fish, fish in enclosures or cages cannot escape from a sudden wave of poisonous or damaging material entering their confined environment. Dinoflagellates have caused massive losses of salmonids in sea cages. The fish may be killed by a nerve toxin produced by some dinoflagellates such as those forming the so-called 'red tides'. Blue-green algae have also been reported as causing losses in fresh water.

'Fish kills' in cages or enclosures are not only caused by direct poisoning but may also be due to suffocation following the removal of all oxygen in the water by the respiration of algae during the night. The physical clogging of the gills of the fish can also cause suffocation, either by the algae or by mucus produced in the gills of the fish as a result of irritation.

Algal blooms can occur at any time in bright spring or summer weather and little is known of the causes. It seems likely however that sudden changes in water temperature or salinity may act as triggers. Nothing can be done to protect the fish in cages or enclosures sited in an area subject to poisonous algal blooms. The only course of action is to move the cages to a safer place.

DISINFECTION

Live fish   It should not be necessary to treat fish for external parasites, if the water supply is free from any organic pollution and saturated with oxygen, provided the fish are not overcrowded and a good standard of hygiene is maintained on the unit.

The condition of the fish and their environment must be taken into consideration, particularly the chemistry of the water. If the oxygen content is low, aeration will be essential during treatment. The chemical used in dips or baths can be more or less active according to the pH, and in acid waters may reach toxic levels even when made up at recommended concentrations. The stress of treatment may itself cause an unjustifiable increase in the casualty rate.

Advice should always be sought if a fish farmer is not completely certain of the diagnosis of the disease and the correct treatment. It is essential to make sure that the fish are not suffering from more than one infestation of parasites at the same time. Gill parasites should always be treated first. As a general rule, the fish should be starved for one or two days prior to treatment as this will reduce the ammonia content of the water.

The strength of the solutions used to make up disinfectant dips and baths should be carefully checked. It is usually too late to do anything about it when the fish begin to show signs of distress.

Dips   The fish are held in a net or sieve and lowered into a concentrated solution, usually for not more than about 30 s.

Baths   The fish are immersed in the chemical solution for up to 1 h (sometimes longer). The oxygen level must be monitored in static water and aeration should be available for use if needed.

When the fish are in tanks or raceways a calculated 'drip' of the chemical at the correct concentration can be added to the water supply. This is expensive but treatment can usually be given in fresh water, without the stress of capturing the fish. Low-level disinfection can be provided over a longer period. Care must be taken with the effluent water containing the chemical solution as this can be a source of pollution.

Formalin   A most useful compound which is commonly used for the chemical treatment of external parasites on fish, this is a 40% solution of pure formaldehyde, uncontaminated by paraldehyde which is poisonous to fish.

Formalin should be carefully handled as it is a respiratory irritant. It must be completely mixed so that the concentration is evenly distributed through the fish tank. In order to make sure that this happens, a little malachite green can be mixed with the formalin as a tracer.

The concentration in the fish tank should not be greater than 1 : 5000 and the concentration should be reduced to 1 : 6000 if the water temperature is over 15°C. The time of treatment should be not more than 1 h.

Formalin reduces the oxygen in the water and aeration should be provided during treatment.

**Disinfection of   Rainbow trout eggs which are imported from any source
eggs**   outside a fish farm should be disinfected on arrival, before being put down to incubate. Packing cases and material should either be burned or disinfected if they are intended for re-use. Similarly, eggs exported from a trout farm to another

unit should be disinfected prior to despatch, and packed in clean material. The normal method of disinfection is to use Acriflavine. A 1:2000 solution, in water containing plenty of oxygen, should be poured over the eggs, in a clean enamel or chemically inert plastic basin. The minimum proportion of Acriflavine solution should be 100 ml/1000 eggs. The eggs should be left in the disinfectant solution for 20−30 min, and the basin or container should be gently agitated to make sure that all the eggs have come into contact with the disinfectant. After treatment, the eggs should be washed in clean water. The Acriflavine should be buffered to a pH 7.7. A correctly made-up stock solution can be obtained from a chemist.

IMMUNIZATION

Methods for the immunizaton against specific diseases so far tried out have been intra-peritoneal injection and hyperosmotic infiltration. The diseases involved are furunculosis and vibriosis respectively. Both methods yield positive results but the difficulty of inoculating large numbers of fish could make this method of immunization a doubtful economic proposition for commercial trout farmers. Commercial vaccines are available for both immersion and injection.

SOME USEFUL CHEMICALS

See Table 12.1.

HYGIENE

The best way to prevent outbreaks of disease on a fish farm is to maintain the highest possible standard of hygiene. This requires that all hatchery equipment, fish troughs, tanks and ponds should be kept as clean as possible under use, and thoroughly cleaned and disinfected at regular intervals. All the general equipment used on a trout farm, such as nets and grading boxes, together with special clothing, rubber aprons

Table 12.1 Some useful chemicals for treating fish

Chemical	Purpose	Method
Benzalkonium chloride (BKC)	Disinfectant	Bath or dip Use depends on pH
Chloramine T (sulphonic acid-free)	Myxobacteria. Gill disease External protozoan parasites	Bath or dip Use depends on pH
Copper sulphate + acetic acid 80%	Bacterial gill disease Fin rot	Bath or dip
Di-*n*-butyl tin (butyl tin oxide)	Flukes and worms Internal parasites	25 mg/kg fish weight for five days
Formaldehyde 40% (Formalin)	External protozoan parasites	Bath or dip
Formalin/malachite green	As formalin	Bath or dip
Iodophors	Disinfection Treatment of eggs	Bath or dip Dependent iodine content
Malachite green	Fungus. *Saprolegnia* Eggs and fish	Bath or dip
Nifurpirinol. Nitrofuran (Furanace. Nifurprazine)	External protozoan parasites Haemorrhagic septicaemias	Bath or dip Oral with feed
Organophosphorous compounds, Nuvan. Neguvon	'Sea lice' *Lepeophtheirus* Caligus	Immersion
Oxolinic acid (Aqualinic powder)	Furunculosis Haemorrhagic septicaemias	50 mg/kg fish weight for 10 days
Oxytetracycline	As for oxolinic acid	50 mg/kg fish weight for 10 days
Sulphamerazine	As for oxolinic acid	200 mg/kg fish weight for 14 days

Note: Treatment with antibiotics in fish feed must cease three weeks before slaughter for market.

and boots, should be frequently disinfected and a disinfectant bath should be maintained for this purpose.

Feed storage   Dry feeds are not dry in the true sense of the word as they have a moisture content of 9–12%. If they are not properly stored or are kept too long in storage, the foods become mouldy and the moulds can produce toxic substances which are poisonous to trout. The vitamins in feeds may deteriorate due to the presence of anti-vitamins in the mixture, or simply as a result of prolonged storage. If the date of preparation is not stamped on bags of feed, this should be obtained from the manufacturer. Feed should be guaranteed free from contamination by the manufacturer, whose responsiblity it is to ensure the good quality of the ingredients. It is, however, as well to inspect any consignment for signs of mould, or other deterioration, before it is fed to the fish. When hand-feeding is used, feed can be hygienically kept in heavy-gauge, plastic dustbins. The bins should be firmly anchored down, with tightly fitting, rain and gale-proof lids.

Hatcheries Egg baskets and trays must be kept free from dead eggs during incubation, otherwise fungus growth will develop and losses may be very heavy. It is better to pick out the dead eggs, using a hand suction bulb or siphon, rather than to use a chemical fungicide such as malachite green. When the fry have hatched and been transferred to feeding tanks, the hatchery troughs and egg baskets should be disinfected and left dry. A cheap solution of potassium permanganate can be used for disinfection. The best modern hatchery equipment is made in grp. It is easy to keep clean, and light enough to be stacked out of the way, when not in use.

The bottom of fry tanks and outlet screens should be kept as clean as possible while in use. After a season's use, the tanks and screens should be thoroughly scrubbed with a proprietary disinfectant and left dry. The insides of fry tanks usually have to be coated with antifouling paint during the off-season, to prevent the growth of algae, and this provides a measure of disinfection. Brushes, or other equipment used in tanks containing fry, should not be allowed to get dirty and should be periodically disinfected.

Cleaning tanks and ponds It is important to keep tanks and concrete ponds as clean as possible. Most tanks with a straight through or circulating current are more or less self-cleansing, but some are less efficient than others. Waste food, faeces, together with any dead fish, must be regularly cleaned out.

Circulating tanks, either rectangular or round with a flat central screen over the drain sump, are generally self-cleansing. Cleansing may become necessary during the first four to five weeks of fry feeding, when the screen mesh has to be kept very small. If cleaning becomes necessary a long-handled nylon broom can be used.

Imported diseases

Economically essential stocking densities and the stressful conditions arising from intensive rainbow trout culture have induced epizootic outbreaks of fish diseases almost unknown and seldom recognized in the wild.

Myxobacteria of the *Cytophaga* spp., such as *Flexibacter columnaris*, resulting in the so-called 'coldwater' diseases, have now become a risk to freshwater rainbow trout culture in the British Isles and in Europe.

The disease known as ERM, or enteric redmouth, caused by the bacterium *Yersinia ruckeri*, which was common in North America has now appeared on British and European fish farms. Rainbow trout infected by these diseases, and some other related fish pathogens, can be treated with oxolinic acid given with their feed.

13 Pollution from fish farms

Causes of pollution

The effluent from a fish farm can have three polluting elements. These are solids in suspension, biological oxygen demand or BOD which is a measure of the pollutants in solution, and residual chemicals from fish disease treatment. The sources of these elements are waste food, faeces, urine and treatment residues such as copper sulphate and formalin from baths as well as fungicides such as malachite green and various antibiotics, including the mycetins and sulphonamides.

The direct effects of the effluent on running waters downstream of a fish farm are to increase the turbidity of the water and reduce the amount of dissolved oxygen. It may also raise the water temperature. Effluents can also be actively poisonous to living organisms, both plants and animals, due to the presence of ammonia, carbon dioxide, nitrogen as nitrite or nitrate and phosphates. Fish farm effluents can also pollute the environment by the release of organisms causing fish diseases, although this is not at present legally regarded as pollution in the ordinary sense of the word.

The changes to the environment caused by fish farm effluents can directly damage the interests of other water users downstream of the farm, such as land farmers and anglers, as well as the interests of other fish farmers. The effect of a polluting effluent on the environment is greatly aggravated if the water in the river into which it has been discharged has not fully recovered its natural chemistry before it reaches the intake of another farm downstream. The cumulative effects of the successive discharges of untreated effluent from a number of farms can be disastrous. The oxygen content may have recovered sufficiently for the fish farms to use the water one after another but the natural clarity of the river water may be spoiled by turbidity from suspended solids.

Pollution load The pollutive strength of the discharge from a fish farm can be directly measured in terms of the solids in suspension and biological oxygen demand or BOD (oxygen taken up in a given period). The statutory limits imposed in Great Britain at present are solids 30 ppm and BOD 20 ppm.

Danger to the environment The polluting matter originating from a fish farm is made up of fish faeces, urine and a little waste food, together with carbon dioxide that has replaced a proportion of the oxygen dissolved in the water. The breakdown compounds that cause damage to the environment are ammonia, nitrogen as nitrate and nitrite, and phosphorus as ortho-phosphate. These can be expressed as nitrogen and soluble, reactive phosphorus in mg/l.

The damaging end-products of farm-fish metabolism depend initially on the coefficient of the feed given to the fish. This is the difference between the total weight of the food eaten by the fish and the proportion that is converted into energy. The coefficients of ordinary fish-feed can range from 1.1 to 1.7 or greater (Table 13.1).

Dangers to other fish farmers The most generally dangerous fish poison in the effluent from any fish farm unit using ponds, tanks, raceways or cages is un-ionized ammonia or UIA. Ammonia is the end product of protein metabolism which in fish is excreted through the gills. If an effluent containing UIA is taken in by another fish farm downstream, recycled on the same farm or released into the water space in which a cage is suspended, it can be directly toxic to fish. Ionized ammonia (NH^{2+}) is not toxic in the amounts present in fish farm effluents.

Table 13.1 Pollution load per tonne of fish production

| Pollutant | Feed coefficient | | | | | |
	1.20	1.30	1.40	1.50	1.60	1.70
Nitrogen (kg)	58.64	57.36	64.08	70.80	77.52	84.24
Phosphorus (kg)	5.88	6.70	7.60	8.50	9.40	10.30

Table 13.2 Ammonia contents. Total ammonia in mg/l needed to produce 0.025 mg/l UIA

Water temp. °C	pH 7.0	pH 7.5	pH 8.0	pH 8.5
5	19.6	6.3	2.0	0.65
10	13.4	4.3	1.37	0.45
15	9.1	2.9	0.93	0.31
25	6.3	2.0	0.65	0.22

A concentration of 0.025 mg/l or more of UIA in the total ammonia load of an effluent is taken as the threshold of toxicity to trout.

The UIA concentration varies in relation to the pH of the water and the water temperature (Table 13.2).

Levels of nitrogen as nitrite between 0.14 and 0.55 mg/l have been found lethal to salmonids but there is no data on toleration.

Carbon dioxide is the end product of respiration. It reduces the oxygen carrying capacity of the fish's blood and is a probable cause, among others, of the condition known as nephrocalcinosis.

Suspended solids causing turbidity reduce the capacity of the fish to feed, extract oxygen from the water and can cause gill damage, particularly to fry.

The chemicals used in the treatment of disease are obviously not lethal to fish when they have been diluted on their return to a water course. The accidental discharge of concentrated chemicals can cause serious losses in another fishery as well as be poisonous to terrestrial animals. Residual antibiotics are dangerous because they may increase the resistance of bacteria pathogenic to other animals as well as fish, including human beings.

Earth ponds Unlined earth ponds provide a certain amount of settlement for solids, and the solids in suspension may be a good deal less than for a corresponding weight of fish kept in tanks. The BOD on the other hand may be higher and additional oxygen is removed from the water by the decaying matter on the bottom of the ponds during the slower interchange of water. Pond liners increase the speed of movement of waste matter through the ponds which then behave more like raceways. The final effluent from earth ponds is difficult to treat to reduce BOD but the total polluting power is less than from tank farms producing the same tonnage. When the ponds are cleaned out, the outlet must be shut off before the ponds are completely empty. The sludge should be pumped into a slurry tank. It makes an excellent manure and should not be wasted.

Tank farms and raceways The BOD is generally less than for earth pond units but the solids in suspension can be very much greater. The interchange of water is too fast to allow waste food and faeces to break down before they are discharged.

Cage farms The main danger comes from the risk of self-pollution by the waste products from cages in fresh water. The risk is not high in the sea unless the cages are anchored in too shallow water or in a place with insufficient current to shift the deposits from below the cages. Self-pollution of cages can be avoided by proper selection of the sites for anchorages.

Treatment The first stage in the treatment of a fish farm effluent must be to remove as much as possible of the solids in suspension. Settlement is the common method but this involves retaining the water for a relatively long period in large ponds. The use of earth ponds can remove the greater part of solids but further settlement may be required. Settlement is usually essential to achieve a satisfactory effluent from a tank or raceway farm. A reduction in the solids in suspension does not mean that the more dangerous ammonia content of the farm effluent has been correspondingly reduced.

Full treatment can be very expensive but the development of the 'separator' (Fig. 13.1) to remove suspended solids combined with the use of an oxygen 'injector' (Fig. 13.2) to restore the dissolved oxygen content of the water and replace most of the gaseous ammonia has gone a long way towards solving the problem. It also allows the effluent water to be

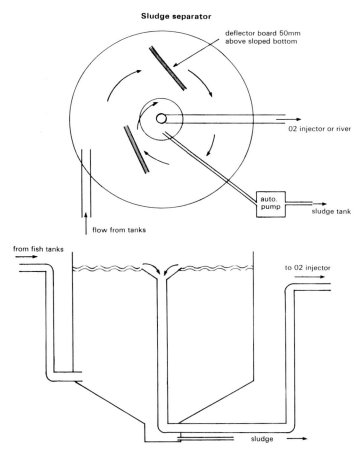

Fig. 13.1 Sludge separator

recycled for use on the same farm which can double production from the available water supply.

The effluent-purifying 'separator' consists of a round tank, 3–6 m in diameter, and approximately 1–2.5 m deep. The bottom of the tank is sloped and forms a funnel into a central sump. The effluent is piped into the tank close to the bottom of the vertical part of the side and is directed round the tank. The heavier particles of waste food and fish faeces move

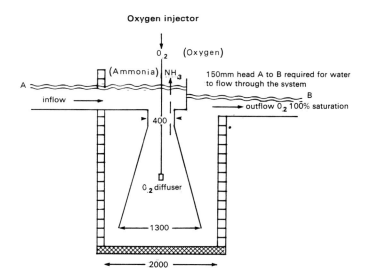

Fig. 13.2 Oxygen injector

downwards in the circulating flow and collect in the sump. The sludge is pumped from the sump at pre-set intervals by an automatic pump (and can be sold as valuable fertilizer to vegetable farmers). The clear part of the effluent, virtually free from suspended solids, rises to the top of the 'separator' where it is piped off. The system, apart from the sludge pump, operates by gravity.

The 'separator' method of treating fish farm effluents is only of use when the fish are reared in tanks. The interchange of water in tanks is sufficiently rapid to keep most of the solids in suspension and carry them out of the tank. The much slower exchange in earth ponds allows settlement to take place and defeats the object of the 'separator'. Most of the solids settle out and collect in the pond where they decompose and further increase the polluting effect of the effluent by increasing the biological oxygen demand.

The effluent flowing from a 'separator', although cleared of solids, is still low in oxygen and may contain too much ammonia. Before being used again to supply another rank of fish tanks (or returned to a river) the oxygen level can be raised and the ammonia removed by an 'injector'. The incoming water passes down a large venturi-shaped pipe into which oxygen is injected through a diffuser linked by a pressure pipe to a standard, interchangeable cylinder. The injection of oxygen has the dual effect of restoring the dissolved oxygen content of the water and driving off most of the residual gaseous ammonia. The flow of water through the 'injector' is entirely by gravity and no pumping is involved at any stage of the operation. The combined effect of a well-designed 'separator' of the corrected size and an oxygen 'injector' can transform the final effluent from a fish farm. The water quality of a river downstream of the outfall can be better than it is at the intake to the farm.

Freshwater fish cages The bed of a lake below a flotilla of trout cages can become foul with accumulated solids from the waste products, and so the cages may have to be moved to allow the bed to lie fallow. The problem is unlikely to arise if the cages are anchored over deep water. The slow descent of solid particles through a water column of more than approximately 30–40 m has been found to have a purifying effect similar to biological filtration. Where cages must be anchored in comparatively shallow water there is a method by which the larger solids in the waste, such as surplus feed pellets, fish faeces and dead fish, can be caught below the cages before they reach the bottom, and removed at intervals.

A reinforced plastic sheet is fitted either inside or attached below the bottom of each net cage. The slope of the sheet funnels waste down to a central sump where a large-bore plastic pipe is set to pass through the sheet and the cage net. The weight of the pipe holds down the sheet and the net. The pipe from below the cage turns up to the surface where it is connected to a pump and the collected waste is then pumped into a tank that can be moved along the walkway. A pump and collecting tank can also be mounted on a service raft and used to clean up a whole flotilla of cages. The solid waste is an excellent fertilizer.

Sea cage
pollution
The percentage of un-ionized ammonia in seawater of a salinity that is low enough for fattening rainbow trout in cages can reach a level that is toxic to the fish. The decomposing waste accumulated below cages anchored in water less than 10 m deep, where the currents are too small to keep the sea-bed clean, could give rise to toxic conditions. Brackish water in the estuaries of rivers may already be polluted to some extent. In seawater un-ionized ammonia is reduced as the salinity increases, but becomes greater as the pH rises (Table 13.3).

pH	Salinity (‰)	Temperature (°C)			
		3	5	7	9
7.5	18–22	0.31	0.36	0.42	0.49
	23–27	0.29	0.34	0.39	0.46
	28–31	0.28	0.33	0.38	0.45
8	18–22	0.98	1.14	1.32	1.92
	23–27	0.96	1.07	1.23	1.43
	28–31	0.90	1.04	1.21	1.40
8.5	18–22	3.05	3.52	4.06	4.69
	23–27	2.85	3.29	3.80	4.39
	28–31	2.79	3.22	3.72	4.29

14 Processing trout

It always pays to put the product as far 'up-market' as possible before it leaves the farm. On-site processing may be an economic necessity if the farm is isolated and transport is difficult and expensive. A great many different things can now be done with the fish to broaden the sales spectrum. They can be gutted and frozen, hot smoked, cold smoked, filleted and packaged. The fillets can be breaded or dipped in batter. Trout can be marinaded, cut in strips and canned like herring. Trout eggs are brined and preserved in glass to be sold as red caviar.

Handling The common mistake made by inexperienced workers is to damage the fish as they are caught up for slaughter. If too many are lifted out at a time they will get bruised and lose scales, which spoils their appearance. Bruised fish become soft and quickly deteriorate. Careful handling is equally important in transporting dead fish from ponds, tanks or cages. If they are dumped into a deep skip, the bottom fish will be half rotten after a short journey in warm weather. They should be packed in proper, shallow fish boxes, and iced if they have to spend more than half an hour in the open air on a hot day. Long distance transport is much better carried out with the fish held alive in tanks and this is the best method to use to bring the fish to a processing centre.

Slaughter All farm fish should be starved before slaughter for long enough to empty the gut. It is common practice, on most trout farms, to let the fish suffocate to death. This may be unavoidable with small or medium size trout but large fish should be deliberately killed as quickly as possible. If the fish die slowly, lactic acids are released into the tissues which accelerate the autolytic process and reduce the length of time the fish can be kept fresh.

The most satisfactory and humane method of killing fish is to introduce CO_2 (carbon dioxide) into the water. The trout have to be confined in a small area in ponds or floating cages and surrounded by a wall of plastic sheeting. The flow can be cut off and the depth reduced in tanks.

Electricity can be used to kill fish that are delivered alive to holding tanks at a processing plant. They can be transferred into a smaller tank where they are killed by 600 v passing through electrodes in the water. It is essential to use a sufficiently high voltage to kill the fish instantaneously otherwise tetanus sets in and vibrates the muscles and ruptures blood vessels, producing black spots of blood that ruin the fish for smoking.

Fish packing in ice

Fish packed in boxes should be covered by flake ice and farms sending fresh trout to market should have an ice maker. Small machines are not expensive and are easy to operate. The quality of the fish is preserved by the ice melting slowly over them which keeps them moist as well as cold. Solid wood or plastic boxes must have holes in the base to allow the melt water to escape.

Large fish should always be cooled on a fish-house floor and washed down with a hose. They should be packed side by side, bellies down in a single layer before being iced.

It is essential to weigh each box of fish accurately before the ice is added. This is vital if the fish are being consigned to a wholesale market or to a commission agent.

Freezing and frozen cold storage

Freezing and frozen storage are by no means the same process. The fish must be frozen in a separate process before they go into deep freeze. What happens in freezing is that the water in the fish is turned to ice by removal of heat. The temperature falls quickly to just below freezing point, then more slowly until about three quarters of the body fluids are turned to ice. The temperature then falls very quickly as there is little heat left to be removed.

The Fahrenheit scale is still used for commercial freezing in English speaking countries, rather than the Celcius (Centigrade) scale. Ice crystals start to form in the fish muscle when the temperature falls below 32°F. Rapid cooling from 32°F down to 23°F results in very small crystals forming inside each tissue cell. The crystals are then not big enough to puncture the cell walls and very little 'drip' or loss body fluid occurs when the fish is defrosted. This is what is meant by the term 'quick-freezing' and is an essential process if the fish are to be properly preserved for cold storage.

The fish in each batch for quick-freezing should be of even size. The temperature can be quickly reduced to 32°F but falls more slowly from 32°F to 23°F (Fig. 14.1). This is called 'the period of thermal arrest' and should be kept to a minimum and should not exceed 2 h.

Slow-freezing has the opposite effect and results in large crystals being formed which break open the cell walls. There is a loss of fluid which gives the flesh a poor texture. The

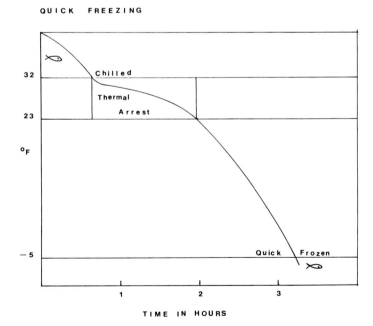

Fig. 14.1 Quick freezing

dehydration which results from the loss of body fluids, con-
centrates the salts and enzymes in the tissues and spoils the
taste of the fish. It also results in a poor quality product if the
defrosted fish are subsequently filleted or smoked.

Rainbow trout are fatty fish and even with properly controlled quick-freezing any tendency to rancidity should be
counteracted by dipping the gutted fish in a solution of ascorbic
acid prior to being frozen.

Types of
freezer
Contact or
plate freezers

The product is sandwiched between hydraulically controlled
plates. The pressure forms frozen blocks and is suitable for
freezing pre-packaged consumer packs. It is not used for high
value fish such as large trout which have to be frozen separately.

Air blast
freezers

The continuous freezer has the product moving steadily through
a tunnel or chamber in a blast of cold air. In the batch freezer
the product remains stationary in the air blast. This is a more
flexible system and caters for a wider range of product shapes
and sizes. Air blast freezers designed for freezing fish such as
large trout individually use a trolley with the fish suspended
singly without touching each other. The type of quick-freezer
to use depends upon the final product, and on the other
processes in the production line such as glazing or packaging
and cold storage.

Packaging
frozen fish

Trout, either whole or fillets, intended for retail sale deep-
frozen in consumer packs should be packaged before quick-
freezing. All fish products deteriorate to some extent in cold
storage due to dehydration and oxidation, particularly fatty
fish. Good packaging in a plasticized carton retards deterioration (Fig. 14.2).

Single, high value fish such as large rainbow trout should
be glazed before going into cold storage by dipping the quick-
frozen fish in water.

Cold storage

If fatty fish are to spend a long time in cold-store they should
be kept at or below $-22°F$ ($-30°C$). If the temperature of the
thickest part of the fish is taken down to $-5°F$ by quick-

Fig. 14.2 Packaging large rainbow trout (*Reproduced courtesy of Danish Seafarmers*)

freezing, the surface temperature will be much lower and the average temperature will be below −20° to −22°F.

A walk-in cold store is a necessity if a quantity of fish is to be kept deep-frozen on the fish farm for any length of time. Ordinary commercial freezers do not keep the fish cold enough for long-term storage, unless they are run at the lowest temperature they can achieve, which is uneconomical for storing large quantities of fish for a long period.

Gutting Trout farmers who do not market their trout through a co-operative are increasingly making use of gutting machines. This reduces waste, improves the keeping time of the fish and gets them to the customer in better condition.

The gutting process involves opening the fish along the centre line of the belly from the throat to the vent. The contents of the body cavity are removed by hand or by a machine. The head is left on but the gills are cut out. The kidney, which lies along the underside of the spine, is brushed-out either by hand or with a mechanical brushing wheel.

Gutting machines come in a variety of shapes and sizes.

The simplest will carry out only some of the operations. Gill-cutting usually has to be performed by hand. The larger machines were originally developed to work at high speed with sea fish. They are expensive to buy and need skilled maintenance, but the best of them will carry out the whole operation and leave the fish clean and ready for quick-freezing or fresh sale.

The type of machine that the farm needs and the speed of operation will depend on the weight of fish that has to be processed. Trout are not harvested at the same rate throughout the year. An expensive gutting machine must have sufficient work to pay for itself. High-output gutting machines may be essential for co-operative processing but even the biggest processing plants handling several thousand tonnes of rainbow trout a year do not operate continuously.

The use of small, simple machines is profitable on most farms producing up to 50 t of trout a year. There is a wide choice but in practice the best is the least complex in design and operation that will do what is wanted. The test of a good gutting machine, large or small, is that it will tolerate minor differences in the size of the fish going through in any batch without adjustment, but is also easily adjustable to take fish in different size ranges. The machine must do a consistently clean job so that the fish do not need hand finishing. Large trout are usually gutted by hand. They vary too much in size to put through a machine.

Filleting   Medium size rainbow trout in the range of 340−454 g (12−16 oz) make excellent fillets. The total weight loss using gutted fish should not be more than 15−20% and there should be no bones left in the flesh. There is at least one American machine designed specifically for trout which turns out a first-class product.

Fillets frozen in consumer packs are one of the best ways to market rainbow trout. The fillets can be factory-coated with crumb or dipped in batter, and the product reaching the consumer tastes as good as it looks.

Smoking   Fish were originally smoked as a means of preservation. Smoking, drying and salting were the only ways to keep fish from going rotten in the days before refrigeration. Trout are now smoked primarily to create a different product. There are two smoking methods, hot and cold. Hot smoking is best used for small fish which are partially cooked in the process. Large fish over 2.3 kg (5 lb) are cut into sides by removing the backbone and smoked at only slightly above the ambient air temperature.

Hot smoking   The fish should first be brined by total immersion in a brine solution at a rate of not more than one volume of fish to three volumes of brine. The temperature should be between 10 and 20°C. The brine must be kept clean and changed when it becomes slimy. The fish can be suspended on 'speats' which are strings through the eyes or tail. The same hanger frames can be used in the smoker. The brining time is shown in Table 14.1.

Table 14.1 Brining time for hot smoking

Weight of trout (*g*)	Time (*h*)
140	2
170	2.5
200	3
250	4
300	5

Hot smoking partially cooks the fish. The 'speats' are loaded into the kiln on frames. The temperature is slowly raised to 86°F (30°C) and held for half-an-hour to give time to dry the skin of the fish. The kiln vents are then closed to keep up the level of humidity and the temperature is raised to 122°F (50°C) for half-an-hour and then to 176°F (80°C) for three-quarters to one hour. The average weight loss in the hot smoking process is 16%.

The smoked fish must be cooled down to below 50°F (10°C)

Fig. 14.3 Smoking kiln (*Reproduced courtesy of Danish Seafarmers*)

and preferably below 39°F (4°C) before packaging. Packages of hot-smoked trout will stay fresh for at least one week if they are cooled down and stored at between 32°F and 39°F (0°C and 4°C). They can be quick-frozen and safely stored for 12 months at −22°F (−30°C).

Cold smoking   The fish must have a fat content of at least 15% of the body weight if they are to smoke well. They will smoke better if they are bled by cutting out the gills at the time of slaughter.

Smaller fish can be prepared for smoking by cutting down from the back to remove the back-bone, then cleaning the fish and leaving two sides joined by the belly.

Large fish should have the body cavity cleaned very thoroughly and the kidney brushed-out. The head is cut off and the fish is then cut down on both sides of the back-bone separating the sides. The lower ribs are left in and the shoulder bones stay on to act as lugs on which to hang the sides during smoking. The sides may be cut from the neck to the tail or stopped at the vent. Sides must be salted or brined before smoking.

Dry salting String loops are put through the shoulder lugs and three cuts made across the width of each side, just penetrating the skin. The sides are laid on a bed of salt and covered with a 1.25 cm (0.5 in) layer at the head and rather less at the tail end. Several sides can be laid on each other with a wooden slat on top carrying a 6.36 kg (14 lb) weight. The sides remain in the salt for 12−14 h according to size. They are then washed to remove the surface salt. Some immerse the sides for about three-quarters of an hour in an 8% brine solution to even out the distribution of salt through the flesh. Brown sugar, molasses and other ingredients are often added to the salt to give a special taste. The loss in weight during dry-salting is about 9%.

Brining The sides are immersed in a 10% brine solution. The brining time is approximately 1 h/0.45 kg (1h/lb) of the average weight of a side. The weight of the sides in a batch should be as even as possible. Molasses, at a rate of about 2.27 kg/45.5 *l* (5 lb/10 gal) and other ingredients, are added to the brining mixture.

Drying Dry-salted or brined sides should be left hanging in cool air for about 12 h. This gives a good gloss on the surface when they are smoked.

Smoking A low, slow fire gives the best results. The heat of the smoke should be not much above the ambient air temperature and not more than 80−85°F (28°C). The smoking time in a forced-draught kiln is 5−10 h according to the size of the sides and the required smoky taste. Some smokers finish off the sides with a short, final burst of hot smoke at about 212°F (100°C) for just long enough to bring the oil to the surface. They believe this improves the appearance of the sides. The loss in weight during smoking is 7−9%.

The very best smoked fish is produced in big, natural-draught smokers, where the fish are lightly smoked at the ambient air temperature for 46−60 h. The best source of smoke is oak shavings from the old sherry casks used for maturing whisky.

Dangers in cold-smoked trout

The form of food poisoning known as botulism resulting from the toxin botulin is fatal. The toxin is produced by a variety of the bacterium known as *Clostridium botulinum*. This organism can be present in the gut of rainbow trout as well as other fish, and also in the mud on the bottom of ponds. It is not the bacterium itself which is deadly, but the toxin that is the product of its metabolism.

Botulin can be destroyed by the temperature and length of time of ordinary cooking. It is food which is eaten raw, or processed and kept in cold store, which can be dangerous. Trout intended for cold smoking should be regularly sampled and tested to make sure they are not carrying the bacteria.

Other methods

Trout can be processed in other ways including canning, but these are factory processes which require the expertise of a specialist in food manufacturing.

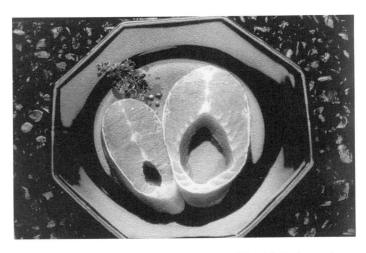

Fig. 14.4 Trout steaks (*Reproduced courtesy of Danish Seafarmers*)

Design and operation of processing plant

The arrangements for processing small quantities of fish present no problems. Common sense is all that is needed to make a convenient layout. The design of a processing line for large-scale production should be the responsibility of a specialist engineer.

Even a small processing plant with a few operatives may be considered to be a factory and have to meet the requirements of legislation passed to protect workers. For example, women employees may need separate lavatories and a rest room. Protective clothing will certainly have to be provided. The working area may have to be heated in winter and cooled in summer. Radiant heaters should be used that warm the people and not the air. Heating pays for itself because cold hands are slow and clumsy.

Fish processing is only worthwhile if it profitably increases sales. Quality control is vital to achieve a high standard in the finished article (Fig. 14.4) and a top-class product deserves a well-designed and attractive package.

15 Markets, profits and losses

The best advice that can be given to aspiring trout farmers is to look for and find a potential market for the fish before starting to farm. They will then at least have some idea how much trout they are likely to sell and the possible selling price, and can see if they are going to make a profit, given the site and the farming method they intend to use.

Modern trout farms are generally specialized in fish production and have to be self-supporting without being able to draw on labour and operating facilities such as storage and transport from outside. They must stand on their own and have to be planned and costed with care if they are to prove successful.

There is no such thing as a truly one-man trout farm because even with maximum automation there are some jobs that always need a spare pair of hands. The highest profit on investment is usually made on a farm producing about 40 t of trout worked by two men as equal partners, with their wives running a farm shop. A small farm will be much more profitable if the fish can be sold locally to hotels and retail shops.

The most economically dangerous size of trout farm is one in the middle range with an annual production of about 50–100 t. This is too much to sell locally and the fish will have to be marketed wholesale but production costs are disproportionately high and very little less than they would be on a farm producing a much larger tonnage of trout. Large farms producing over 200 t of fish a year have a progressively increasing economy of scale.

Profit margin Trout farming is correctly regarded as a high risk enterprise. It is not unusual to have a substantial loss of stock every four or five years and a quick return on investment is essential. A well-sited and well-run unit with good marketing should return between 35% and 40% profit on turnover before tax, and pay off the invested capital in less than three years.

The elements making up the cost account of a trout farm vary not only in relation to the location of the farm in a particular country but between one country and another. Interest on capital may be much the same but the cost of fish food and labour will be different. Farms sited close to markets may sell their trout at a better price but may have to pay more in wages. Isolated units must have the lowest production costs to pay for the long distance transport of fish food and fish.

Like any other industry the profitability of trout farming depends on keeping down costs. Maximum productivity is usually achieved on the largest farms where 100 t a man year is not unusual with automation. Small farms with working owner—managers depend for their profits on the low-cost labour and a disregard for the overtime that has to be worked every day of the week, throughout the year.

The capital investment and running costs per tonne of annual production differ according to the design of the farm. The most common cause of financial disaster in large, mechanized farms is that they have been grossly over-capitalized and actual production never reaches the design capacity.

Running costs of trout farms	Interest on capital
	Depreciation
Fixed costs	Rent and rates
	Labour and management
	Insurance
	Other charges
Variable costs	Fish food
	Repairs and maintenance
	Transport
	Electricity or other power
	Disposable equipment
	Security
	Losses of fish
	Dry and cold storage

Percentages of running costs The cost of fish food expressed as a percentage of the total running costs (Table 15.1) is a measure of the economic efficiency of a trout farm. The feed cost can range from 40% of the total on less efficient to 65% on the most efficient farms.

Table 15.1 Percentages of running costs

	Pond farm	Tank farm	Cage farm
Food	65	62	63
Labour	15	14	18
Others	20	24	19

Growth rate If the fish take longer than expected to reach market size and do not do well, the fault may be in the water quality or temperature but is more likely to be due to bad husbandry and incorrect feeding. Food is by far the highest production cost. Commercial pellet feeds have a standard conversion rate. If this is not being met or bettered the farm will run into financial difficulties sooner or later. There is very little margin for error and it is essential to keep an accurate check on food quantity against growth rate.

Trout for the market The preferred size for table market trout has changed little in recent years. 'Portion' fish weighing 170−227 g (6−8 oz) for fresh sale are still in demand. Larger fish of 340−400 g (12−14 oz) fetch a better price because they can be processed and sold as deep frozen fillets in consumer packs. The small fish are still used by the 'hot' smokers but big fish of 1.8−2.75 kg (4−6 lb) are needed for 'cold' smoking for sale in competition with smoked salmon.

Short-falls in production Bad planning and managerial over-optimism combined with lack of experience are the usual causes of failure in meeting production targets. Bad management or unskilled husbandry must be detected and corrected. It is never possible to profit by mistakes in any kind of commercial aquaculture.

The root cause of failure may be inherent in local conditions. The wrong conclusions can have been drawn from the original survey or the selected farming system prove incorrect for the site. Under these circumstances the position may be irretrievable.

Fish losses   Trout farms are continuously at risk. Fish are dying all the time, sometimes for no apparent reason. Human error is the commonest cause of attributable loss. Failure of the power supply and mechanical breakdown of pumps or other machinery, the failure of water channels, pipelines and dams due to faulty construction or lack of maintenance can nearly always be avoided by taking proper precautions. Loss due to predators, humans, birds and mammals, can be prevented. Loss prevention is a matter of constant vigilance by everyone working on the trout farm. A trained fish master, like any other stockman, should be able to see at once if his fish are not doing well. There are some unavoidable dangers which can lead to large-scale losses.

Floods   Ordinary flooding can be avoided by proper site selection but no-one can prevent an occasional disaster.

Drought   Water supply is the first and principal consideration in aquaculture. The quality and quantity must be as secure as humanly possible but there may then be the hottest summer in living memory.

Storm damage   This is a problem for cage farmers. It is seldom possible to find a site sheltered from all quarters. A strong gale of wind can come from a totally unexpected direction.

Pollution   A milk factory upstream may 'accidentally' discharge ammonia into the water. A tanker truck carrying acid can run through a bridge parapet into the river just above the intake.

Disease   Your friendly, neighbouring fish farmer can pass on his IPN or something worse. Your next consignment of fingerlings may be infected before they are delivered.

Insurance The high risk in fish farming means that insurance premiums charged for protection against stock losses are correspondingly high. A 'franchise' under which losses have to reach an agreed percentage of the total value of the fish held on the farm before a claim can be made is one way to reduce the premium. Another method is called 'deductible' which means that the fish farmer agrees to bear the first part of any claim.

Many fish farmers only insure against disasters, because it would cost too much to cover all everyday risks. A trout farmer faced with the problem of the cost of insurance must decide where he is most vulnerable to a loss which he cannot afford to bear and then to insure against those specific risks. Insurance is most vital when the stock on the farm has reached its maximum value. It may be possible to negotiate a policy in which the premium is calculated on a variable month by month basis according to the estimated value of the fish stocked at that time.

Specialist production Farmers specializing in the production of eyed eggs and others in the sale of fingerlings from their own brood stock are an established part of the trout farming industry. Specialization requires a much higher degree of expertise than ordinary table-market farming. It also requires a site with a clean and chemically infallible water supply at the right temperature. The majority of highly infectious diseases are at their worst in juvenile fish and scrupulous attention to hygiene is essential. Good brood stock management needs a degree of sympathy with the animals which is not easy to achieve with fish. Successful specialist producers have usually had a long apprenticeship in aquaculture and also have an interest in their work which goes beyond financial reward. This is the most interesting branch of the trout farming industry but it is also the most difficult and demanding of skilled fish husbandry.

Co-operatives Very large farms can be better off on their own. Small farms may get a higher price for their fish through semi-retail outlets. It is the medium size farms producing 50–150 t that stand to gain most by forming a co-operative. Individually they do not produce enough fish to command a stable price but together their combined production can be enough to influence the overall market.

The medium-size trout farmer on his own is at the mercy of the wholesaler or has to sell his fish to a factory unit for processing. A well-run co-operative should be large enough to carry out its own processing and operate a cold store. The members then get the profit from the increased sale price of their own fish rather than profiting others who have no hand in the hard business of raw food production.

Is it worthwhile? The initial decision to go ahead can only be made if the whole project is viable in detail and in depth. The following list is an indication of various pieces of essential information that should be obtained and examined.

Location Access. Transport costs. Proximity to markets.

Water supply Availability. Quantity. Chemistry. Temperature. Cost.

Site Basic cost. Selection of production method. Processing on site. Development cost.

Product Species of fish. Size at sale. Time taken to reach saleable size.

Capital Funds available. Contingencies. Inflation. Interest. Independence.

Construction costs Water supply. Hatchery. Fry tanks. Fish tanks. Cages. Ponds. Jetties. Processing buildings and stores. Roads. Fences.

Working capital Estimated date of profitability.

Estimated production costs Starting with eggs or buying fingerlings. Food. Estimated live weight gain. Fish processing. Transport. Insurance. Disease treatment.

Estimated fixed costs Heat, light and power. Maintenance. Equipment. Rent. Rates. Water supply.

Labour   Availability at local rates. Estimated man hours per ton of annual production. Cost per man hour.

Salaries   Owner. Manager. Pension funds.

Sales and office   Advertising. Insurance. Postage. Telephone. Travel.

Loan charges   Interest on bank or private loans. Depreciation.

Now, having weighed up the pros and cons, if you decide to become a fish farmer, may the best of luck go with your venture.

Index

acclimatisation (sea), 95
acidity (water), 93, 142
Acriflavine, 131
aeration, 55, 92
Aeromonas (disease), 115
alevins, 32, 37
algae, 128
amino acids (diet), 102
ammonia (UIA), 137, 141, 142
anaesthetics, 27
anchorages, 71, 81
Arctic char, 6
automatic feeding, 97, 108

bacterial diseases, 114, 134
baskets (hatching), 30
BKD (disease), 116
BOD (pollution), 136
botulism, 152
brood stock, 20, 21
brown trout, 2
buying eggs, 23

cage farms, 138
    design, 72
    mesh, 73
    servicing, 73
    sites, 71
cages
    freshwater, 71, 141
    sea, 81, 82, 141
calories (diet), 103
carbohydrates (diet), 101
carbon dioxide, 137, 144
carotenoids (feed), 109
chemicals, 132
cleaning (tanks, ponds), 55, 133
coefficients (feed), 106
cold smoking, 150, 151
cold storage, 144, 146
co-operatives, 159

costiasis (disease), 118
counting
    eggs, 32
    fry, 95
cross breeding, 24
cryogenic storage (eggs), 22
Cytophaga (myxobacteria), 134

Danish system, 12, 45
densities (fish), 60, 67
diagnosis (disease), 127
diet
    deficiencies, 110
    fish, 101
dinoflagellates, 128
disease risk, 61
diseases, 113, 134
disinfection, 56, 129, 133
'dry' feed, 104

earth ponds, 12, 137
    aeration, 555
    cleaning, 55
    construction, 46
    disinfection, 27, 56
    effluent, 57
    feeding, 50
    grading, 51
    'monks', 49
    screens, 49
egg production, 19
    counting, 32
    disease free, 23
    fertilisation, 25
    size, 21
    stripping, 24
electric killing, 144
enclosures (shore), 18, 80
equipment, 92, 94
ERM (disease), 134
eyefluke (disease), 122

failures (causes), 157
farm sizes, 155
fats (diet), 102
fattening fish (sea), 85
feed (fish), 105
    cost, 111
    storage, 132
feeding times, 108
    in cages, 74
fertilisation, 25, 74
fish losses, 158
flesh colour, 109
floating 'islands', 84
food conversion ratio, 112
formalin, 130
freezers, 128
freezing, 144
fry production, 38
    density, 41, 42
    flows, 41
    grading, 38
    indoors, 45
    tank covers, 43
    tanks, 39
    water supply, 40
fungicide, 31
fungus (fish), 128
furunculosis (disease), 114

genetics (broodstock), 22
gill disease, 116
glazing (quick freezing), 146
grading, 38, 51, 74
growing time, 108
growth control, 111
growth rate, 157
gutting (processing), 147
Gyrodactylus (parasite), 121

handling (fish), 97, 143
hatcheries, 28, 34
hatching, 32
heated water, 96
heat exchangers, 97
hexamitaisis (disease), 119
husbandry (fish), 85, 91

hybrids, 24
hygiene, 131

incubation (eggs), 30
induced sterility, 23
IHN (disease), 125
immunisation, 131
'injector' (oxygen), 140
insurance, 159
IPN (disease), 124

maintenance, 85
market sizes (fish), 157
minerals (diet), 102
moist pellets, 106
'monks', 49

nets, 73
nitrogen (pollution), 112, 136

octomitis (disease), 119
osmosis, 76, 86
oxolinic acid, 134
oxygen, 76, 141

packaging (fish), 146
pellets commercial, 105
    hygroscopic, 107
    moist, 46
phosphorus (pollution), 112, 136
pollution, 112, 135
    dangers, 136
    sources, 138
    treatment, 138
pond culture, 46
    aeration, 55
    cleaning, 55
    disinfection, 27, 56
    effluent, 57
    feeding, 50
    grading, 51
    screens, 49
'portion' trout, 57, 157
processing, 153
production load (pollution), 136
profitability, 160
profit margins, 155

protein
    diet, 101
    pollution, 112

quick freezing, 145

raceways, 13
    construction, 57
    disease risk, 61
    feeding, 60
    fish density, 60
    flow, 60
    grading, 60
    investment, 61
rainbow trout, 1
re-circulated water, 88
recycling
    closed, 99
    low-head, 68
running costs, 156, 157

saltwater tolerance, 77
sea anchorages, 81
sea cages, 81
'sea lice' (parasites), 127
seatrout, 5
seawater shore-farms, 78
screens, 49, 93
scuba diving, 85
security, 98
sex reversal, 22
sexual maturity, 109
shore bases, 73, 81

shore-walkway cages, 83
silage feed, 106
sites, 71, 92
slaughter (fish), 53, 75, 143
'steelhead' trout, 1, 86
sterility (induced), 23
stocking, 85
stripping broodfish, 24
supersaturation, 33

tanks, 14, 63
    construction, 64, 67
    density (fish), 65
    flows, 67
    materials, 64, 65
    solar heating, 70
    water supply, 65, 67
tidal enclosures, 80
troughs (hatchery), 29

VHS (disease), 125
viability (projects), 92
vibriosis (disease), 115
vitamins (diet), 103, 110

water supply, 9
    chemistry, 9
    temperature, 11, 76
whirling disease, 117
white spot disease, 119

Yersinia ruckeri (bacterium), 134

Books published by **Fishing News Books**

Free catalogue available on request from Fishing News Books, Blackwell Science, Osney Mead, Oxford OX2 0EL, England

Abalone farming
Abalone of the world
Advances in fish science and technology
Aquaculture and the environment
Aquaculture & water resource management
Aquaculture development – progress and prospects
Aquaculture: principles and practices
Aquaculture in Taiwan
Aquaculture systems
Aquaculture training manual
Aquatic ecology
Aquatic microbiology
Aquatic weed control
Atlantic salmon: its future
The Atlantic salmon: natural history etc.
Bacterial diseases of fish
Better angling with simple science
Bioeconomic analysis of fisheries
British freshwater fishes
Broodstock management and egg and larval quality
Business management in fisheries and aquaculture
Cage aquaculture
Calculations for fishing gear designs
Carp farming
Carp and pond fish culture
Catch effort sampling strategies
Commercial fishing methods
Common fisheries policy
Control of fish quality
Crab and lobster fishing
The crayfish
Crustacean farming
Culture of bivalve molluscs
Design of small fishing vessels
Developments in electric fishing
Developments in fisheries research in Scotland
Dynamics of marine ecosystems
Ecology of fresh waters
The economics of salmon aquaculture
The edible crab and its fishery in British waters
Eel culture
Engineering, economics and fisheries management
The European fishing handbook 1993–94
FAO catalogue of fishing gear designs
FAO catalogue of small scale fishing gear
Fibre ropes for fishing gear
Fish catching methods of the world
Fisheries biology, assessment and management
Fisheries oceanography and ecology
Fisheries of Australia
Fisheries sonar
Fishermen's handbook
Fisherman's workbook
Fishery development experiences
Fishery products and processing
Fishing and stock fluctuations
Fishing boats and their equipment
Fishing boats of the world 1
Fishing boats of the world 2
Fishing boats of the world 3
Fishing ports and markets
Fishing with electricity
Fishing with light
Freshwater fisheries management
Fundamentals of aquatic ecology
Glossary of UK fishing gear terms
Handbook of trout and salmon diseases
A history of marine fish culture in Europe and North America
How to make and set nets

The Icelandic fisheries
Inland aquaculture development handbook
Intensive fish farming
Introduction to fishery by-products
The law of aquaculture: the law relating to the farming of fish and shellfish in Great Britain
A living from lobsters
Longlining
Making and managing a trout lake
Managerial effectiveness in fisheries and aquaculture
Marine climate, weather and fisheries
Marine fish behaviour in capture and abundance estimation
Marine ecosystems behaviour & management
Marketing: a practical guide for fish farmers
Marketing in fisheries and aquaculture
Mending of fishing nets
Modern deep sea trawling gear
More Scottish fishing craft and their work
Multilingual dictionary of fish and fish products
Multilingual dictionary of fishing vessels/safety on board
Multilingual dictionary of fishing gear
Multilingual illustrated dictionary of aquatic animals & plants
Multilingual illustrated guide to the world's commercial coldwater fish
Multilingual illustrated guide to the world's commercial warmwater fish
Navigation primer for fishermen
Netting materials for fishing gear
Net work exercises
Ocean forum
Pair trawling and pair seining
Pelagic and semi-pelagic trawling gear
Pelagic fish: the resource and its exploitation
Penaeid shrimps — their biology and management
Planning of aquaculture development
Pollution and freshwater fish
Purse seining manual
Recent advances in aquaculture IV
Recent advances in aquaculture V
Refrigeration on fishing vessels
Rehabilitation of freshwater fisheries
The rivers handbook, volume 1
The rivers handbook, volume 2
Salmon and trout farming in Norway
Salmon aquaculture
Salmon farming handbook
Salmon in the sea/new enhancement strategies
Scallop and queen fisheries in the British Isles
Scallop farming
Seafood science and technology
Seine fishing
Shrimp capture and culture fisheries of the US
Spiny lobster management
Squid jigging from small boats
Stability and trim of fishing vessels and other small ships
The state of the marine environment
Stock assessment in inland fisheries
Study of the sea
Sublethal and chronic toxic effects of pollution on freshwater fish
Textbook of fish culture
Trends in fish utilization
Trends in ichthyology
Trout farming handbook
Tuna fishing with pole and line